MICROBIC DISSOCIATION III
FILTRABLE BACTERIA

AND
FURTHER ADVANCES IN THE STUDY OF
MICROBIC DISSOCIATION

Philip Hadley,
Edna Delves and John Klimek

Hygenic Laborator, University of Michigan, Ann Arbor (1931)

AND

Philip Hadley

Institute of Pathology, Western Pennsylvania Hospital, Pittsburgh (1937)

*[Reprinted from Journal of Infectious Diseases,
Vol.48, No.1 (1931) and Vol.60, No 2 (1937)]*

I0478328

With a FOREWORD by
S. H. SHAKMAN
InstituteOfScience.com
Santa Monica, CA USA
2013

Published in the USA
by the Institute Of Science

FOREWORD
S. H. Shakman

WHAT IS A VIRUS, REALLY?

In 1931, when Philip Hadley etal. published the article "Filtrable Bacteria",
its brief 2-word title itself was the accepted and virtually complete definition
of "viruses". More than a decade later the situation was unchanged, as stated
in the authoritative 1944 *Encyclopedia Britannica*:

> "VIRUS, see FILTER-PASSING-VIRUSES"
> "FILTER-PASSING VIRUSES, organisms small enough to pass through a
> bacterial filter made of unglazed porcelain or compressed infusorial earth.
> In these filters the grains of the china clay or infusorial earth used for their
> manufacture are sufficiently small and uniform to leave interstices, the
> cross-section of which is 0.2 to 0.8 u in diameter. If a liquid containing
> microbes whose [p. 239] smallest diameter exceeds 0.2 u be pressed
> through the filter, the microbes remain impacted in the smaller crevices. As
> 0.2 u is also the limit of size of a particle which can be resolved by the best
> microscopes (see microscopy) when white light is used, filter-passing
> organisms are either invisible or on the margin of visibility. Hence most of
> them have been classed as "ultravisible viruses".

That's it! As of the World War II era, the sole distinguishing difference
between a bacterium and a virus was size. But as of a couple of decades
later, as per the 1966 *Britannica*, this had been modified:

> "VIRUSES, once usually referred to as filterable viruses, may be defined as
> self-reproducing agents smaller than the microscopically visible bacteria,
> multiplying only within living susceptible cells and responsible or
> potentially responsible for a wide range of infectious diseases. ...
> "After 1930 there was a steady development of new technical methods
> for the physiochemical and biological properties which completely changed
> the outlook of biologists toward the viruses. Probably the single most
> important advance toward this new approach was W. J. Elfords
> development of "graded" filter membranes of known pore size. This
> allowed for the first time an opportunity of assessing the actual sizes of the
> particles responsible ...
> "... Then in 1949 tissue culture methods long used in half-hearted
> fashion for the cultivation of viruses were shown by J. F. Enders and
> colleagues to be suitable for the growth of poliomyelitis viruses. This not
> only opened a way to immunization against polio, but supplied a method by
> which dozens of new types of virus were isolated."

Thus, notwithstanding Britannica's 1966 indication of a supposedly
"completely changed" biologists' view of the virus, the declaration that

improved filter technology was "the single most important advance" since 1930 nonetheless seemingly reaffirmed size as the ultimate determinant of whether an entity was viral or bacterial. But how (or if) this "most important" advance might explain the (new) definition of a virus (as "multiplying only within susceptible cells"), or how it might otherwise be explained, was not clearly laid out.

Presumably the explanation underlies the brief discussion of the Enders etal. culture method breakthrough, which was an advance on Sabin and Olitsky's successful cultivation of poliomyelitis virus in human neural tissue (the product of which tended to be very toxic); Enders etal. used non-neural embryonic tissue, and the realization that this virus cultivation was possible, without the use of neural tissue, enabled the requisite mass production associated with Salk and Sabin poliomyelitis vaccines.

OK, the work of Enders, as well as that of Sabin and Olitsky before them, did substantiate that virus *can be produced within living cells*; and without spinning off on a tangent that could itself spawn a series of volumes, so do a number of other instances (e.g. influenza) demonstrate that viruses *can be* produced within living cells. But is it thereby established that viruses *can only be produced in this manner*? Absolutely not.

Indeed, the central premise of this 3-part series of compilations, showcasing the works of Philip Hadley and associates, addresses this very point, central to an understanding of the nature of bacteria and viruses. This series exhaustively documents how viruses can and do multiply without the vehicle of a "susceptible cell".

As Hadley had discussed in his 1927 historical review (pp. 8-9 of the first volume in this 3-volume series), the idea of microbial variability was "crystallized" in the work of Nageli in 1877; however, the renowned, even revered, Emile Koch took the contrary position of fixed bacterial types advocated by the botanist Cohn in 1875, and through the force of Koch's prominence the Cohn-Koch view became established "dogma" in the late 19th Century. Notwithstanding the works of Baerthlein, Arkwright and others in the early 20th Century that thoroughly documented the phenomenon of dissociation (so much so that Hadley in the 1920s-1930s was seemingly convinced that the world must, indeed would, come to its senses and embrace it), the rigidity of the Koch view has fundamentally persisted even up to modern times. Certainly we do find discussions of variability (i.e. mutations and even dissociations) in the literature, but these seem to be treated as special circumstances rather than a general quality that inhabits microorganisms of all types.

Specifically regarding the organism associated with poliomyelitis, as early as 1932 E.C. Rosenow, Head of the Division of Experimental Bacteriology of the Mayo Foundation for three decades (1915-1944), in association with Rife and Kendall at Northwestern University, demonstrated the existence of filter-passing forms (phases) of the streptococcus isolated from poliomyelitis. In 1935 Rosenow reported on the cultivation of sub-microscopal forms, with a new methodology involving chick mash cultures, that for the first time enabled "production with considerable regularity the symptoms and lesions of poliomyelitis and encephalomyelitis ..." Rosenow would later demonstrate that the filterable entity thus produced from the implicated streptococcus and the natural virus of poliomyelitis and other conditions were indeed indistinguishable from each other.

[Rosenow, E.C., Proc. Mayo Found., July 13, 1932. 28; Science, Aug. 26, 1932; Proc. Mayo Found., June 19,1935.]

Rosenow's filterable poliomyelitis microbe was not cultivated in human tissue, or any living tissue for that matter. Rather it was produced by very specific culture methods, developed and proved over the prior 3 decades by Rosenow at Mayo, from the streptococcal phase of the organism. Nonetheless, the filterable phase of this organism as cultivated by Rosenow is indeed the very same as that referenced and utilized by Enders, Sabin, and Salk. Dissociation from streptococcal to the viral form and reconstitution of the streptococcal was well-documented by numerous tests. A brief bibliography of some of Rosenow's subsequent definitive studies on the relation between the streptococcal and viral phases of various diseases, with emphasis on poliomyelitis and influenza:

Rosenow, E.C.:
--- Microdiplococci in filtrates of natural and experimental poliomyelitic virus compared under electron and light microscopes, Proc. Staff Meet., Mayo Clin. 17: 99-106, Feb. 18, 1942.
 --- Relation of neurotropic streptococci to encephalitis and encephalitic virus, Proc. Staff Meet., Mayo Clin. 17: 551-560, Nov. 4, 1942.
--- Studies on the relation of a neurotropic streptococcus and virus to epizootic encephalitis of wild ducks, Cornell Veterinarian 33: 277-304, 1943.
--- Production of filtrable infectious agent from alpha streptococci, Am. J. Clin. Path. 14: 150-167, March 1944.
--- Poliomyelitis; relation of neurotropic streptococci to epidemic and experimental poliomyelitis and poliomyelitic virus, diagnostic serologic tests and serum treatment, Internat. Bull. Econ. M. Research and Pub. Hyg. A44: 9-83, 1944.
--- Studies on virus nature of infectious agent obtained from 4 strains of "neurotropic" alpha streptococci, J. Nerv. and Ment. Dis. 100: 229-262, Sept. 1944.

--- Filterable infectious agent obtained from alpha streptococci isolated in studies of case of poliomyelitis, Am. J. Clin. Path. 14: 519-533, Oct. 1944.
--- Studies on relation of pneumotropic streptococci to influenza virus, Science 100: 434-435, Nov. 10 1944.
--- Streptococci and diplostreptococci and respective "viruses" in etiology and epidemiology of epidemic respiratory infections and infectious gastroenteritis, Am. J. Digest. Dis. 17: 261-270, Aug. 1950.
--- Streptococci, diplostreptococci and respective filtrable agents from outdoor air during epidemics of respiratory infections and infectious gastroenteritis, Am. J. Digest. dis. 18: 155-163, May 1951.
--- Parallel production of altered infectivity of Streptococcus and related filtrable agents isolated from outdoor air, J. Aviation Med. 22: 225-243, June 1951.

Yet however interesting might be the science demonstrating transitions between streptococcal and viral phases of these pathogens, perhaps even more significantly (from a very practical perspective), as reported by Rappaport in 1948, a therapeutic poliomyelitis vaccine regimen developed by Rosenow was successfully used during the 1946 polio epidemic on 20 polio patients who had already showed signs of poliomyelitis infection. All had fever and some combination of positive Brudzinski, stiff neck, absent reflexes, nausea, and/or vomiting, etc. Fifteen of the 20 left the hospital with no residual weakness, and the remainder had greatly improved. And successful treatment of an additional 20 children during the epidemic of 1948 was reported in 1954; all 20 walked out of the hospital, in stark contrast to the many who were otherwise condemned to the dreaded iron lung or death (vaccination also was 100% successful prophylactically on family members of the children successfully treated).

Rappaport, Benjamin, *Journal-Lancet,* October 1948, p.395-7; *Quarterly Bulletin, Northwestern University Medical School* **28**, 1954, p. 57-60;

This therapeutic use of a poliomyelitis vaccine was unprecedented, and seemingly nothing even remotely similar has ever been accomplished in any sense by any investigator or team. The clear and universal assumption, even today, is that once infected with poliomyelitis, whatever other methods of therapy might be employed following the crisis, there is no such thing as a therapeutic polio vaccine. The fact that 40 children were infected and stricken by various degrees of poliomyelitis paralysis, which infection was for the most part aborted through use of this vaccine; virtually all of which children, without this regimen, would have worsened and many would have died or ended up in iron lungs; it is rather incredible that 35 of the 40 were promptly restored to health and the remainder were much improved and improving.

The point here of course is that even the year prior to the highly touted work of Enders et al in 1949, which enabled the cultivation of the viral phase of the poliomyelitis organism; Rosenow's methodology, using antibody derived from the streptococcal phase, was performing miraculous reversals in children already stricken with polio, and it was 100% successful prophylactically in limited tests involving family members. But notwithstanding this admittedly obscure success, the world would have to wait nearly a decade longer for a purely viral-phase based prophylactic poliomyelitis vaccine.

The work of Tulasne (and others) further substantiates that multiplication may occur in the absence of "susceptible [host] cells", e.g., as a consequence of penicillin as per Tulasne, or otherwise as discussed by Hadley, e.g. that

Tulasne R, Nature 164 (Nov. 19, 1949)

Hence a bacterium as common as *Proteus* gives, under the action of penicillin, dwarf, submicroscopical, 'filtrable' forms with possible reversion from the dwarf to the normal form (see diagram).

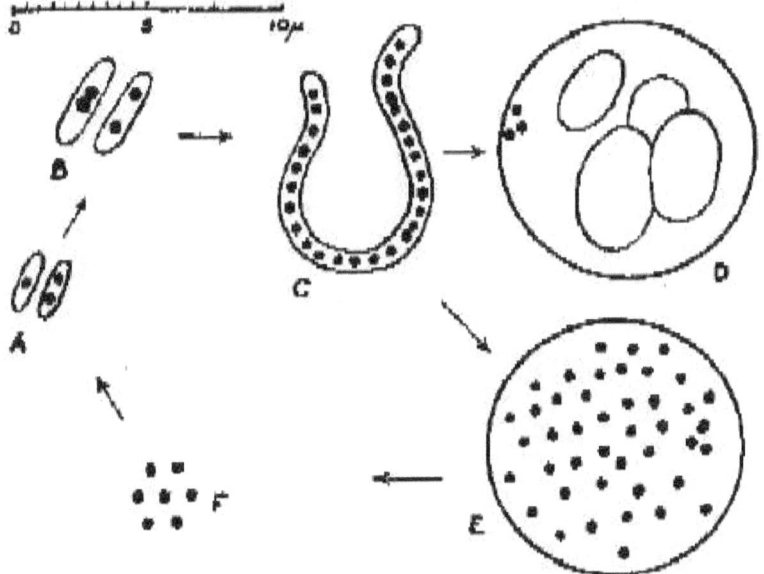

A, Normal *Proteus* cells, resting phase; B, normal *Proteus* cells, lag phase; C, swarming filament; D, large body in lysis; E, large body with desoxyribonucleical granulations; F, pleuropneumonia-like body (desoxyribonucleical granulations)

bacteriophage [auto-]generation (with the possibly-mandatory "liberating influence" is a form of reproduction. Tulanse's diagram of the dissociation of proteus under the influence of penicillin as illustrative of dissociation and replication within and by proteus itself.

Thus, rather than representing an "advance" as suggested by the 1966 Encyclopedia Britannica, the supposition that viruses are capable of "multiplying *only* within living susceptible cells" may comprise a fundamental flaw and constraint in our current understanding of the true nature of viruses. And without a correct understanding of viruses and their relation to bacteria, we will be forever without an understanding of the nature of human disease.

S. H. Shakman
Institute of Science
Santa Monica, CA, USA
InstituteOfScience.com
I-o-S.org
25 July 2013

CONTENTS

THE FILTRABLE FORMS OF BACTERIA: A FILTRABLE STAGE IN THE LIFE HISTORY OF THE SHIGA DYSENTERY BACILLUS*

Philip Hadley, Edna Delves and John Klimek**

Hygenic Laboratory of the University of Michigan, Ann Arbor, Mich.
Jour. Infect. Dis., Vol. 48, No. 1 (1931)
Submitted for Publication, Aug. 12. 1930

AND

FURTHER ADVANCES IN THE STUDY OF MICROBIC DISSOCIATON

Philip Hadley

Institute of Pathology, Western Pennsylvania Hospital, Pittsburgh, Pennsylvania, USA
Jour. Infect. Dis., Vol. 60, No. 2 (1937)

---

*The present paper constitutes the third in a series dealing with the subject of bacterial variability. The first (Microbic Dissociation) was published in this journal in 1927. The second (The Twort-d'Herelle Phenomenon) was published in this journal in 1928. The present report, which continues this study of dissociative variations among bacteria, contains a statement regarding the method of production, the cultural behavior, the morphology and the filterability of what may be termed "the filterable virus stage" of the Shiga dysentery bacillus.

**On the junior authors has fallen the task of much exacting laboratory work in conection with the hundreds of filtrations and plate cultivations involved in this study. For the plan of the experiments and for the conclusions advanced, the senior author alone is responsible.

FURTHER ADVANCES IN THE STUDY OF MICROBIC DISSOCIATION (1937)
PHILIP HADLEY

INTRODUCTION

It can scarcely be doubted that significant discoveries in bacteriology have often resulted from focusing attention on the unexpected or apparently irrelevant, just as suggestive paths into new fields of research have often been passed by because of the assurance of the investigator regarding the exclusive nature of what he was about to discover—or regarding what he believed was impossible of discovery.

To visualize too much truth in apparently inconsequential details is, no doubt, a hindrance to the development of a science. Moreover, it is one thing to suspect a new truth and another to demonstrate it to the satisfaction of one's colleagues. But the exercise of some imaginative faculty, as Johannes Müller once remarked, is no less necessary for the scientist than for the poet. In the realm of bacteriology it is perhaps possible that a certain blindness to the unexpected, and a certain inelasticity in evaluating possibilities, have done, and are doing, much to fetter the advance of this science into new fields. This is most noticeable whenever the dissociative aspects of bacterial behavior are under discussion; and it can be observed particularly whenever the "filtrable forms of bacteria" or "bacterial life cycles" are in question. This field of research is admittedly hazardous—for experiment, conclusions or controversy. Moreover, it has come about in these days that to express convictions that differ from the *consensus gentium* becomes almost professional foolhardiness; it brings down the strictures of one's friends and enemies alike. In the presentation of this paper we are not unmindful of this circumstance. But we are also conscious of the fact that, beneath the tumult of controversy between monomorphism and pleomorphism, there is being born a new epoch in bacteriology, the limits of the significance of which and the possible future expansion of which no one can yet surmise.

Incontrovertible proof of the existence, in the developmental history of common bacterial species, of a distinctly filtrable, but at the same time cultivable, stage or stages, differing markedly from the forms of culture already commonly known to bacteriologists, would possess significance from several points of view. First, it might give more complete knowledge of the biology of the species and of its specific cyclogeny. Second, it might afford a criterion for the recognition, not of apparent, but of actual, sterility of bacterial cultures and animal tissues, of toxins and immune serums. Third, it might yield an entirely new picture of the extent of the distribution of pathogenic microbes through the organs, blood and other tissues of the infected animal or man, and thus permit deeper insight into certain problems of pathology. Finally, such proof might afford a new basis for the study of the still obscure etiology of several important communicable diseases that at present appear to be caused by filter-passing micro-organisms.

In view of these important possible eventualities, provided success was attained, it seemed desirable to attack in a very definite way and, as will later be seen, from a somewhat new point of view, the much disputed subject of the existence of filtrable forms of bacteria, attempting to relate, if possible, the phenomena observed to dissociative behavior.

The early studies of Novy and Knapp on the filtrability of the spirochete of relapsing fever, the results of which were later confirmed by Breinl and Kinghorn, Nicolle and Blanc, Wolbach and others; of Lourens on the hog cholera bacillus; of Fontès on the tubercle bacillus; of Almquist (1911) and Friedberger and Meissner on the typhoid bacillus; of Hort (1916, a) on the meningococcus; of Löhnis on *Azotobacter,* and Miehe and Mellon (1920, 1926) on various species—to mention only a few of the older reports—have possessed considerable significance in the minds of unprejudiced workers who have not been too closely bound by the dictates of monomorphism, although it is true, some of the investigators named themselves saw nothing noteworthy in their results. On the other hand, the point of view maintained in some of these earlier works, particularly those of Leishmann, Löhnis, Miehe and Mellon, has received substantial support through the results and conclusions of more recent contributions in this field. These embrace especially the studies of Calmette and his school, Vaudremer, Hauduroy (1929, e), Togounoff, Valtis, Mellon and Jost and, most recently, Panek and Zakharoff, on the tubercle bacillus; of Morin and Valtis on the bacillus of Johne's disease; of Fejgin (1925) and Hauduroy (1926, a) on the typhoid bacillus; of d'Herelle (1922) and d'Herelle and Hauduroy on the Shiga bacillus, the cholera vibrio and the staphylococcus under the influence of bacteriophage; of Pryer, Ramsin, Ramsin and Givikovitch, Sédaillan and Gaumond, Rosenow, Evans, Zlatogoroff (1927), Urbain, Hauduroy (1926, b, 1927, a) and Hauduroy and Lesbre, on the streptococcus; of Hauduroy (1927, b) on the diphtheria bacillus [1]; of Burnet (1926) on *Bacillus pestis;* of Haag on *B. anthracis;* of Fejgin (1924) on the X19 strain of *B. proteus;* of Levaditi (1929) on the Dutton spirochete; of Esther Stearn and her collaborators on an organism from tumors; of Séguin on *Treponema calligyrum* and *Spirochaeta gallinarum,* and of Manouélian on the spirochete of syphilis.

But the significance of these often mutually confirmatory reports has, for the most part, remained unappreciated by the majority of bacteriologists. Against the view of the existence of special, filtrable forms of bacteria and spirochetes have been urged many and diverse arguments by writers who have never taken the trouble seriously to study the question or who have not applied the proper methods. Other workers, for example, Burnet (1928) and Palante and Koudriavtzeva, while admitting the possible existence of such filtrable forms, have sounded a note of warning, especially with reference to the so-called "microbes de sortie" (when animal tests for the demonstration of the virus forms have been employed). Bronfenbrenner took a stand against the existence of special filtrable forms, as did Frobisher and Cooper and Petroff, chiefly as a consequence of negative results in the attempted filtration of the tubercle bacillus. Mudd likewise doubted the existence of "filtrable phases" of bacterial growth; and this represents the most common textbook position on the subject at present. Arkwright, in an excellent review of variation phenomena in bacteria, stated: "Ultramicroscopic forms of pathogenic bacteria are not generally accepted by bacteriologists as a definite phase or stage in their life

---

1. After this paper was ready for the press there appeared the report by Smith and Jordan dealing with the filtrable form of Bacillus diphtheriae. In eleven of nineteen old cultures, these forms were demonstrated in Chamberland filtrates as a result of subjecting them to the method of serial agar plate cultivation as recommended by Hauduroy. From five to nine passages were usually required. In the development of the original culture type out of the so-called "protobacterial" forms, the microscopic elements underwent a sequence of morphologic changes: "a granular stage, followed by the appearance of giant cocci, which are replaced by micro- and diplococci, and ultimately by pleomorphic bacillary forms." A study conducted by Miss Richardson and one of us (P. H.) during the past year or more on the G forms of the Park 8 strain confirms in nearly all respects the results reported by Smith and Jordan. The only difference lies in the circumstance that we produced the G forms before filtration was attempted, by subjecting the original culture to the influence of lithium chloride for the purpose of forcing the dissociation into the filtrable stage. Our complete results will be presented at a later date. Certain aspects are embodied in a later section of this work.

history." One suspects that Arkwright himself does not regard the issue as settled. Indeed, notwithstanding the results of Vaudremer, of Hauduroy (1929) and of Panek and Zakharoff in the cultivation from filtrates of the tubercle bacillus, the majority of workers still oppose the view that special, filtrable forms of this organism exist.

On the other hand, the existence of a few filter-passing organisms which are, at some time in their life, cultivable in or on artificial mediums has been recognized for some years. Thus we have *Asterococcus mycoides* of contagious pleuropneumonia of cattle, manifestly filtrable, but at the same time cultivable, as demonstrated by Bordet, by Nocard and Roux, by Borrel, by Oeskov and by others; also the virus of aphthous fever, cultivated by several workers; also *Bacterium pneumosintes,* studied by Olitsky and Gates, Neufeld (1928) and others. Regarding the filtrable forms of common bacteria, however, there has been much doubt, although the truth of the matter has been strongly suggested by a sufficient number of investigations.

When these various and conflicting reports of the filtrability of bacteria are examined carefully, and the experimental conditions reviewed critically, flaws can be detected in certain cases; sometimes proper controls have not been set up or maintained. Moreover, as one of us (P. H., 1927) pointed out, an uncertainty is likely to arise in any case in which animal inoculations have been employed to demonstrate the presence of living elements in the filtrates, as, for example, in the work on the tubercle bacillus. On the other hand, in the face of so many studies alleging the discovery of filtrable forms in so many species and by so many qualified investigators, it might appear that the results obtained and the conclusions reached could not be destitute of significance in every instance. One feature of all the earlier studies is, however, somewhat remarkable—that the reader is so often left to believe that the elements which passed the filter candles were similar to those of the visible culture before the filtration was performed. The significance of this point will become clearer in later pages of this paper.

There are, of course, many ways of viewing these reported successful results. In general, however, they have been regarded as accidental; and in the same light, indeed, have they often been regarded by the authors themselves. Seldom has it been accepted that the elements that passed the candles were special cell forms which, by reason of their size, form or charge, were susceptible to filtration. Aside from the claim that the results were accidents, the "contamination hypothesis" has most commonly been brought against the reported results, as indicated in an article by Frobisher, to be considered later.

But, suppose for a moment, that, instead of denying the existence of special, filtrable forms of bacteria on general principles, as often seems to be the case, or on the grounds of a preconceived monomorphism, which is worse, we attempt to picture to ourselves what might be some of the actual conditions surrounding the generation of such bodies, if they exist; and what might be their behavior in attempted cultivation in or on common culture mediums. In the first place, let us suppose that these filtrable bodies are not present at all times in every bacterial culture, but that they arise only at certain moments in the course of culture development. This is no more than we should expect from present knowledge of the behavior of the S, O and R forms of culture in microbic dissociation. Let us assume, moreover, that such filtrable bodies might not always be cultivable, at least at once, on any culture medium, solid or liquid; or that they might grow at the expense of one, but not of the other. If there is any truth in these surmises, our experimental efforts to detect the filtrable elements should be directed accordingly. In the first place, we should not attempt to discover them in every culture, nor in a given culture at all times. In the second place, if the filtrable bodies should be present in numbers only at a certain stage of culture development, it might happen that their manner of manifestation would be different from that of culture types with which we are familiar. Moreover, the detection and propagation of such forms might demand methods somewhat different from those commonly employed for the detection and cultivation of common bacterial species. It might be that, for a time, like the seeds of plants, they would not visibly develop.

There is still another matter, of even greater importance, that might have a bearing on the best technic for demonstrating the existence of these filtrable bodies. It relates to the following circumstance. The only distinct "culture types" at present recognized for any bacterial species are the so-called S, R and O forms; and the first generation and interconvertibility of these types have been proved to stand in intimate relation with the phenomena of microbic dissociation. It seems probable that, if there exists a hitherto unrecognized culture type different from these and characterized by ready filtrability and perhaps ultramicroscopic dimensions, such a type would also stand in some sort of relation to the dissociative reaction, and in harmony with the laws of this reaction, so far as they are known, would appear only at a certain point of culture development and then vanish from the culture. Such a mode of behavior can be observed in any dissociating culture with reference to the S and the R forms, and constitutes the basis of a hypothesis that can justifiably be opposed to the common, but unfounded, one of the identity of the filtrable bodies with minute, bacterial "fragments," such as the "arthrobacteria" of Burnet (1926), or perhaps with the "infrabacteria" of Nicolle (1925).

Somewhat radical considerations of this sort, probably meaningless to those who have not enjoyed personal laboratory experience with dissociative reactions in their favorite cultures, have led us to investigate the problem of the existence of filtrable forms of bacteria by what we believe to be a distinctly new method of attack: namely, to discover the first origin of these elements in relation to the forced dissociative behavior of the mother culture; and then to study them as representing a transient, but perhaps highly significant, cyclostage in the developmental history of the bacterial species. The degree to which this new point of view (and therefore new method of attack) has been successful will be revealed by the following experimental report. It is accordingly the aim in the present paper to present the chief facts regarding an investigation, conducted during the past three years or more, on the mode of origin and behavior of a filtrable stage that has been produced in the life history of the Shiga dysentery bacillus; also to record, although without further description at the present time, the existence of analogous forms of culture in *B. coli*, *B. typhosus*, *B. paratyphosus* A and B, *B. enteritidis*, *B. cholerae-suis*, *B. typhi-murium* I, *B. typhi-murium* II (*B. pestis-caviae*), *B. diphtheriae* (Park 8 strain), *B. acidophilus* and *Vibrio cholerae*, observed and to some extent studied either by ourselves or by other members of our laboratory group.

In this study we have had one further aim, but one that will receive scant mention in the following pages—to throw light on the nature and mechanism of bacteriophage action. We have not dealt particularly with the bacteriophage in this investigation, for it is sufficient to attack one phase of the problem at a time. But, to the critical reader, it will perhaps become apparent that we have been dealing with certain complex phenomena that lie behind it—phenomena on the further interpretation and explanation of which, we are confident, the mystery of the bacteriophage will sometime be solved.

We freely admit that the problem of the filtrable forms of bacteria is surrounded by numerous uncertainties, and that a given chain of experimental evidence may receive various interpretations. We have, moreover, recognized the great opportunity for the creeping in of unsuspected agencies for error in our conclusions. For these reasons we have attempted, from the outset of our work in 1927, to fortify our methods at those points that seemed the most vulnerable by a series of control tests which may appear to those who have not had experience in such fields of study as somewhat elaborate. These tests have been incorporated in a separate section.

ORIGIN OF G TYPE CULTURES AND OF ASSOCIATED FILTRABLE FORMS

Since the beginning of this study the filtrable virus stage of the Shiga dysentery bacillus has arisen more than sixteen times during the spontaneous or enforced dissociations of either the S or the R form of the Shiga organism. In only one instance did it arise in a culture in which the dissociative changes were not being followed at the time. This filtrable form, for reasons that will later become apparent, will hereafter be referred to as the G type culture.

The spontaneous dissociations preceding or accompanying the generation of the filtrable form of culture occurred, and were studied, in plain beef infusion broth having a reaction of from p_H 6.8 to p_H 7.8. The enforced dissociations were produced under the influence of from 0.25 to 0.5 per cent lithium chloride beef infusion broth having a reaction of from p_H 7.5 to p_H 7.8, or under the influence of a 5 per cent pancreatin (Squibb's) beef infusion broth having a reaction of p_H 6.8 or p_H 7.8. In other instances, the dissociative reactions were produced under the influence of rabbit or guinea-pig peritoneal exudate removed some hours after the injection of large volumes of bacterial suspension. In order, however, to avert possible criticism relating to the use of material from the animal body as a dissociating agent, it may be said at this time that the results obtained by this method will receive brief further mention; for our conclusions are in no way dependent on these data. We believe, however, that the results are of confirmatory significance. In addition, they touch an important phase of our problem which cannot receive further treatment in this paper—namely, the rôle of body fluids in determining the production of filtrable bodies from the recognized culture types.

The dissociative reactions in the experiments about to be reported were produced in part by serial transfers through the medium of choice. Much of the beef infusion broth employed was given a reaction of p_H 7.8, because it is now well known that an alkaline medium usually favors the S to R transformation. It may easily accomplish such changes even without the addition of foreign substances. In some cases, the serial transfers were started with an S type Shiga culture. In others, an R strain was employed, earlier produced in the laboratory from known S stock. The amount of culture transferred from tube to tube varied from one loop (standard 4 mm.) to several drops (30 drops = 1 cc.). Each broth tube in the series of transfers, after from twenty-four to forty-eight hours' incubation at 37 C., was spread, in proper dilution and by means of a sterile, bent glass rod, over the surface of a sterile agar plate previously prepared. The plates were incubated for from twenty-four to ninety-six hours or more and then examined for variations in colony form (differential colony count). It was by this method that the changing colony picture accompanying the dissociative reaction could best be followed. Of course, the most common colony picture was that of varying proportions of S, SR and R colonies, one type giving place to another with the continued transfers. At intervals, and usually preceded by a rather characteristic appearance of the colonies, colonies that we have termed the G type appeared. In fact, the G type culture was first detected in colony form. These minute and often microscopic colonies, a description of which will be given later, were therefore first observed in definite relation to the changing proportions of S, SR or R colonies, occurring about the same time in the same culture, and often revealed on the same agar plate. The serial transfers in plain or in modified broth were usually continued well beyond the point in the series at which the G colonies first appeared. This was for the purpose of ascertaining their further behavior in relation to the other cyclostages present in the ever changing bacterial population.

The G form of the Shiga cultures has been observed more than sixteen times, and has been produced, isolated and cultivated separately on fifteen occasions and by a variety of methods.[2] In some cases, this culture form remained stable for a considerable time, especially in sealed

2. These instances do not include the numerous occasions on which the analogous form of culture has been obtained by similar methods for other members of the colon-typhoid-dysentery group and several other bacterial species.

ampules. In others, it reverted, after continued transfer, to the original type of culture. As will be mentioned later (section on reversion), the degree of stability of the G type cultures seemed to depend, in a considerable degree, on whether a filtration through a Berkefeld N or W candle was interposed in their early history and soon after their first isolation. After the following general account of the origin of these cultures, a more detailed statement of the origin of a few selected strains will be given.

BRIEF STATEMENT REGARDING ORIGIN

Strain I of the G cultures arose as follows: A typical S type Shiga culture, isolated from the laboratory stock, was transformed to an R form by preliminary serial passage through 0.5 per cent lithium chloride broth having a reaction of p_H 7.8. This R culture was then passed serially, at twenty-four hour intervals, through 0.25 per cent lithium chloride broth having a reaction of p_H 7.5. The G culture arose in colony form on the agar plate streaked from the tenth broth tube in the series (table 1). The number of G colonies on this plate was roughly 4,000.

Strain Ia of the G cultures came from the same R type culture used in the aforementioned experiment. In this case, however, the R culture was passed at twenty-four hour intervals through plain infusion broth of reaction 7.5. The G colonies first appeared on the plate streaked from the tenth tube in the series (table 1). The number of colonies on first appearance was roughly 5,000.

Strain II arose from an S type Shiga culture undergoing passages at twenty-four hour intervals through 0.25 per cent lithium chloride broth of reaction 7.5. The G colonies first appeared on the plate streaked from the sixth tube of the series. The first indication of dissociative action in the culture of passage occurred in the broth tube immediately preceding (table 2).

Strain IIa arose from the same S type culture undergoing serial passages at twenty-four hour intervals through plain infusion broth of reaction 7.5. The G colonies first appeared on the plate streaked from the seventh tube of the series (table 2) and their appearance immediately followed the first indication of dissociative action (tube 6) observed in this series. The close parallel shown in the origins of strains II and IIa is well shown in table 2.

Strain III arose from an R type culture undergoing serial passages at twenty-four hour intervals through 5 per cent pancreatin broth of reaction 7.8. The first G colonies arose on the plate streaked from the twelfth serial tube (table 3) immediately following a "sterile" plate made from the eleventh broth tube.

Strain IV arose from an S type culture undergoing serial passages at twenty-four hour intervals through 5 per cent pancreatin broth of reaction 7.8. The first G colonies arose on the plate streaked from the twelfth broth tube in the series (table 4). The preceding plate made from the eleventh tube in the series was "sterile." The similarity in the conditions surrounding the origins of strains III and IV will be observed from the data presented in tables 3 and 4.

Strain V arose from an S type Shiga culture undergoing serial passages at twenty-four hour intervals through 5 per cent pancreatin broth of reaction 6.8. The first G colonies appeared on the plate from the seventh tube in the series, following a sterile plate from the preceding (sixth) tube in the series (table 5). No G forms arose in a similar broth series with pancreatin omitted.

Strain VI appeared on an agar plate that had been streaked with some S type broth culture in addition to a small amount of the alpha fraction of the homologous Shiga bacteriophage. In other words, the G type colonies, in this instance, arose as secondary resistants to the lytic agent. They seemed to be identical with the G forms obtained by other means not involving intentional use of bacteriophage. These G type colonies furthermore showed marked resemblance to the fine, secondary colonies that often arise on lytic areas (plaques) caused by the action of a properly diluted lytic filtrate on a background of sensitive culture.

Strain VII was obtained from a sealed (in glass) tube containing a Berkefeld filtrate of lytic agent that had been preserved in this manner for several months at room temperature and in the dark. Plates streaked with a few drops of the filtrate showed, after several days, the fine growth of the G type colonies. Similar results were obtained from several other tubes of lytic filtrate that had been stored from one to six months and that included strains active on both the Shiga bacillus and *B. typhosus.*

Strain VIII was produced as follows: Into the peritoneal cavity of a rabbit was injected about 25 cc. of a twenty-four hour broth culture of the type S Shiga stock. After ten hours, 25 cc. of sterile broth was injected by the same route, and the animal was chloroformed at once. A quantity of cloudy peritoneal exudate was then removed aseptically by means of a sterile Pasteur pipet and the full amount brought to 35 cc. by the addition of fresh, sterile broth. The mixture was then filtered through a Berkefeld N candle under 200 mm. of negative pressure and collected in a sterile, cotton-plugged tube. Of this filtrate, 10 drops were then spread over the surface of a sterile, plain agar plate, which was incubated for twenty-four hours. No sign of colonies appeared; the plate seemed sterile. The plate surface was then washed over with a few drops of sterile broth and the "washings" transferred to a second sterile plate. After an incubation of twenty-four hours or more, this plate also appeared sterile. The process was repeated, and the "washings" were transferred to a third plate, which was incubated for twenty-four hours. At the end of this time, the surface was studded with multitudes of minute colonies measuring from 0.05 to 0.1 mm. in diameter. From one of these colonies a G type culture was derived by subculture to plain infusion broth. Similar results were obtained with the S type of *B. typhosus* inoculated in a similar manner.

Strain IX had the following history: On April 24 there was made an agar slant culture of a pure line S growth. On May 14, the culture was transferred to plain infusion broth and incubated for forty-eight hours. At this time there was observed a fair, homogeneous clouding resembling that of the "normal" S type culture. On May 16, this broth culture was streaked on agar and revealed, on May 17, three kinds of colonies—S, SR and G. These colonies appeared in increasing numbers according to the order indicated. The G colonies were subcultured to plain infusion broth, yielding a pure line G strain.

Strain X arose from an S type Shiga culture undergoing serial passages at twenty-four hour intervals through plain infusion broth having a reaction of 7.8. The G colonies first appeared on the plate from the twelfth serial tube.

Strains XI and XII were obtained from samples of Shiga bacteriophage (717 and 927) produced in the laboratory for this purpose. The G type cultures were isolated from the apparently sterile filtrates, after aging, by the serial plate method. The origin of one of these strains (XI) will be presented in detail on a later page.

Strain XIII arose in an aged broth tube of an S type Shiga culture. After growth for about twenty-four hours at 37 C., the tube was placed in the cold room at a temperature of 8 C., where it was kept for two years. At the end of this period, it was cultured to agar, but no growth resulted. The tube was then incubated for two weeks at 37 C. At the end of this time, cultivation on an agar plate yielded numerous G colonies, with colonies of another type resembling the R, but much smaller. From one of the G colonies the type G culture was obtained.

Strain XIV also arose in a tube of normal infusion broth culture left to age at room temperature. This culture made from S type growth was examined at intervals for fifty-eight days. The G colonies were observed in the plates made on the eleventh, sixteenth and twenty-second days. They were absent before the eleventh day and after the twenty-second, the observations being discontinued on the fifty-eighth day. The detailed records of this experiment are presented later in this section.

Strain XV was produced in July, 1930, by dissociating an S type Shiga culture in beef infusion broth, p_H 7.8, containing 0.35 per cent of lithium chloride. The G colonies first appeared on the litmus lactose agar plate spread with culture from the sixth serial tube. The G colonies were accompanied by S, and a few SR, forms.

Other instances of the production of the G type culture from either the S or the R form of the Shiga bacillus involved merely duplications or slight modifications of the methods thus far presented. It early became apparent that the generation of the G forms was not dependent on the presence of lithium chloride, pancreatin or bacteriophage in the culture medium, but resulted in many instances in mediums made up only of alkaline infusion broth. It is important to state that these results were never obtained in acid mediums.

It should also be noted that the manner of producing these G forms appeared to make no difference in the character of the cultures obtained. There are many methods of enforcing the dissociative reaction on a sensitive culture, and we believe that the same results might be obtained by methods quite different from those mentioned in this paper. In fact, Plantureux obtained marked dissociations, also artificial generation of bacteriophage, by the use of the chlorides of calcium, barium and strontium in his alkaline culture mediums. It seem probable that a homologous immune serum effects the same end.

As adding significance to the results of our work with the G type culture of the Shiga bacillus, it should be added that both ourselves and other workers in this laboratory, attempting to confirm or extend our observations, have succeeded in obtaining analogous strains of G type cultures in the following bacterial species: *B. coli* (two instances); *B. typhosus* (seven instances); *B. paratyphosus* A (four instances); *B. paratyphosus* B (two instances); *B. enteritidis* (one instance); *B.*

typhi-murium (two instances); *B. typhi-murium* II (one instance); *B. cholerae-suis* (one instance); *B. acidophilus* (one instance); *V. cholerae* (one instance), and *B. diphtheriae* (two instances)   Only three attempts to produce the G forms have met with failure thus far; these involved *B. psittacosis*, *Bact. tumefaciens* and *B. carotovorus*. In every instance, moreover, the G forms produced were demonstrated to be filtrable through N and W candles—or, at least, to have associated with them filtrable bodies.   The methods used were, for the most part, the same as those outlined in foregoing paragraphs.

DETAILED STATEMENT REGARDING ORIGIN OF SELECTED STRAINS

In view of the important circumstance that, in the majority of the instances noted in the foregoing paragraphs, the origin of the G cultures was intimately associated with the process of active dissociation in the mother culture, it seems desirable at this point to supplement the brief statements already made with further details in a few cases.   This will have the advantage of presenting a picture of the manner of appearance of the G colonies in relation to the ever changing proportions of S, SR and R colonies, with which the G forms were associated in the tubes and plates.   The strains selected for this purpose are the following: I, Ia, II, IIa, III, IV, V and XIV.   The data concerning the mode of origin of these strains may be accepted as indicating the general trend in the generation of analogous strains produced and studied by the same methods.

Dissociations Forced by Lithium Chloride.—Origin of Strains I and Ia: The history of the culture from which strains I and Ia were derived was as follows:

In November, 1927, a pure line S type Shiga culture was passed at intervals of twenty-four hours through broth tubes containing 0.5 per cent lithium chloride. Dissociation into the R form occurred in the fourth serial transfer. A typical, large, irregular, translucent and spreading colony was selected and tested for R purity by plating.   This plate, yielding 100 per cent R colonies, gave the R culture employed in the experiment.

Tubes of beef infusion broth having a reaction of p_H 7.5 and containing 0.25 per cent of lithium chloride (0.5 per cent is likely to cause too great an inhibition of growth) were prepared; also a similar series of plain broth without the lithium chloride.   The purity of the R stock culture having been verified again, one standard loop (4 mm.) of broth culture was inoculated into the first tube of each series.   The tubes were then incubated for twenty-four hours at 37 C., after which spread agar plates were prepared for each tube and inoculated with appropriate dilutions.   The plates were then incubated for from twenty-four to forty-eight hours and the colony forms examined and recorded. After this the plates were kept under observation for from four to six days longer. Just before plating, transfers were made to the second tube of each series. At the end of twenty-four hours, the second pair of tubes were plated and transfers made to the third pair; and so on through a series of more than fifty transfers. At each transfer, a record was made of the degree of growth in broth and the kind of growth (i. e. homogeneous or agglutinative); also regarding the presence or absence of sediment. Table 1 presents a record of these transfers, and the appearance of the colonies revealed from plating each tube in the series. The figures in the columns headed S, SR, R, G, etc., represent the number of observed colonies of the different types on the basis of 200 colonies examined. In some cases, the exact proportions were not recorded and then the number is merely indicated as "many" or "few." The number of G colonies, when these were present, is usually reported outside of the 200, on which the counts of the other colonies are based.

The significant facts brought out in table 1 are, first, that the G type colonies, characterized by their pinpoint size and having an average diameter of from 0.1 to 0.2 mm., first appeared to the unaided eye in the plates spread from the tenth broth tubes in both series, namely, the lithium chloride and the plain broth. In the plain broth series, after persisting for some days, they disappeared again (sixteenth day), being replaced (seventeenth day) by S type colonies. On the twenty-first and twenty-third days they again appeared in small numbers, but disappeared again on the twenty-fourth day and were not observed again during the following twenty-six days, at the end of which the series was terminated.

In the lithium chloride broth series, however, the G type culture persisted, off and on, from the time of its apparent origin in the tenth tube to the end of the series. The S type colonies were observed for the last time in the nineteenth plate and the SR and R colonies were last observed in the twenty-first. From this time on only the G colonies remained. During this period many "sterile plates" were recorded, alternating with plates that showed only the G colonies. The significance of these appearances and disappearances is not at present clear, but the subject will be mentioned later in connection with certain phenomena observed in the bacteriophage reaction. Some of the "sterile plates," after standing in the laboratory for from five to ten days, showed a small number of G colonies.[3]

---

3. In the records of the first appearance of the G colonies here presented, it will be understood that they were always detected by the unaided eye. At this point in the study, we had not learned that colonies might be present so minute that they could be seen only by the aid of a hand lens or the microscope.

TABLE 1.—*The Origin of Strains I and Ia of the Shiga G Type Culture and Their Relation to the Accompanying Cyclostages*

| | Strain I: Lithium Chloride Broth | | | | | | Strain Ia: Broth without Lithium Chloride | | | | |
| | Number of Colonies | | | | | | Number of Colonies | | | | |
Tube*	S	SR	R	G	Remarks	Tube*	S	SR	R	G	Remarks
1	1	...	199	.....		1	0	..	200	.....	
2	2	...	198	.....		2	0	...	200	.....	
3	20	...	180	.....		3	2	...	198	.....	
4	106	...	94	.....		4	20	...	180	....	
5	114	...	86	.....		5	16	...	184	.....	
6	104	...	96	.....		6	36	...	164	.....	
7	128	...	72	.....		7	166	...	34	.....	
8	154	...	46	.....		8	193	...	7	.....	
9	138	...	62	.....		9	199	...	1	.....	Reversion
10	174	...	26	4,000	First G	10	1	...	199	5,000	First G
11	178	...	22	2,000		11	0	...	72¶	128¶	
12	160	40	0	1,000		12	0	...	120¶	80¶	
13	188	12	0	6		13	Few	26	0	174¶	
14	0	13†	0	?	Plate nearly sterile	14	Few	?	Few	Many	
15	178	Few	22	Few		15	0	Few	0	Many	
16	198	0	2	Few		16	0	0	0	0	Sterile
17	198	0	2	Few		17	200	0	0	0	
18	199	0	1	0		18	120	70	10	0	
19	170	0	30	0		19	110	60	30	0	
20	0	0	0	0	Sterile	20	60	140	0	0	
21	0	192	8	Few	S type disappears	21	0	0	1†	197†	
22	0	0	2†	Many		22	200	0	0	0	
23	0	0	0	3,000		23	200	0	0	Few	
24	0	0	0	200		24	200	0	0	0	G forms disappear
25	0	0	0	0	Sterile	25	1†	1†	0	0	
26	0	0	0	Many		26	2	198	0	0	
27	0	0	0	0	Sterile	27	0	120	80	0	
28	0	0	0	Many		28	20	140	140	0	
29	0	0	0	0	Sterile	29	20	40	140	0	
30	0	0	0	Many		30	4	0	196	0	
31	0	0	0	0	Sterile	31	1	0	199	0	
32	0	0	0	Few		32	0	0	200	0	
33	0	0	0	0	Sterile	33	0	0	200	0	
34	0	0	0	10,000		34	0	0	200	0	
35	0	0	0	0	Sterile	35	0	5	195	0	
36	0	0	0	10,000		36	0	0	200	0	
37	0	0	9	50		37	0	0	200	0	
38	0	0	0	500		38	0	20	180	0	
39	0	0	0	1,000		39	0	0	200	0	⎫
40	0	0	0	2,000	Some 0.5 mm.	40	0	0	200	0	⎪
41	0	0	0	2,000		41	0	0	200	0	⎪
42	0	0	0	1,000		42	0	0	200	0	⎬ Stabilized R
43	0	0	0	Few		43	0	0	200	0	⎪
44	0	0	0	Few		44	0	2	198	0	⎪
45	0	0	0	?	All 0.5 mm.	45	0	0	200	0	⎪
46	0	0	0	Few		46	0	0	200	0	⎭
47	0	0	0	Few		47	1†	0	3†	0	⎫
48	0	0	0	Few		48	48	100	52	0	⎬ Partial reversion to S
49	0	0	0	Few		49	20	76	104	0	⎪
50	0?	0	0	0?	Mostly 1 mm. G or S (?)	50	50	110	40	0	⎭

* Serial transfers at twenty-four hour intervals in 0.25 per cent lithium chloride broth (strain I) and plain infusion broth (strain Ia), of reaction 7.5.

† Gives actual count; in other cases, the figures are given on the basis of 200 colonies. The usual number of colonies counted was from 100 to 200.

¶ G colonies figured in the 200 count.

Origin of Strains II and IIa: Running parallel with the tests just recorded concerning the R type Shiga culture were other experiments in which the S type was employed. The culture was an S stock that had been purified by repeated platings and could be depended on, at the moment when the experiment was started, to yield 100 per cent S colonies. This culture was transferred in series, at twenty-four hour intervals, through 0.25 per cent lithium chloride broth, a control culture being passed at the same time through plain infusion broth. Both mediums had a reaction of 7.5. Platings from each tube of the series, after twenty-four hours' growth, were carried out as in the experiment

TABLE 2.—*The Origin of Strains II and IIa of the Shiga G Type Culture and Their Relation to the Accompanying Cyclostages*

	Strain II: Lithium Chloride Broth						Strain IIa: Broth without Lithium Chloride				
	Number of Colonies						Number of Colonies				
Tube*	S	SR	R	G	Remarks	Tube*	S	SR	R	G	Remarks
1	200	0	0	0		1	200	0	0	0	
2	200	0	0	0		2	200	0	0	0	
3	200	0	0	0	Stable S	3	200	0	0	0	Stable S
4	200	0	0	0		4	199	0	1	0	
5	84	116	0	0	First reaction	5	200	0	0	0	First reaction
6	148	52	0	65	First G	6	85	115	0	0	First G
7	188	12	0	60		7	200	0	0	10	
8	200	0	0	1,000		8	200	0	0	100	
9	200	0	0	2,000		9	200	0	0	107	
10	200	0	0	Many		10	200	0	0	Few	
11	200	0	0	200		11	200	0	0	Few	
12	0	0	0	0	Sterile plate Terminated	12	200	0	0	Few	
						13	0	0	0	Many	
						14	0	0	0	Many	
						20	200†	0	0	?	Colony size increases
						25	200†	0	0	0	
						36	200†	0	0	0	
						38	200†	0	0	0	

* Serial transfers of an S type Shiga culture at twenty-four hour intervals through 0.25 per cent lithium chloride broth (strain II) and plain infusion broth (strain IIa) of reaction 7.5.
† From this point on, the colony size increased by slow degrees to 1, 1.5 and then 2 mm., the colonies coming to resemble the S type. At a certain stage, the colonies might have been regarded either as G or as S.

just described. The results of these two series, which are typical of others, are presented in table 2.

From the data presented in table 2 it appears that, with the S type culture being used as a source, the G colonies first appeared in the lithium broth series in the sixth serial tube, while in the plain broth series they appeared first in the seventh serial tube. In this experiment, as in the one previously described (table 1), it was demonstrated how similar experimental conditions serve to cause the liberation of the G forms at about the same moment. In the test previously detailed, they arose in both series in the tenth tubes of transfer. The same situation was observed in several of the experiments which are to be described later. In addition, we have on record two instances dealing with *B. typhosus* in which the tenth serial tube represented the "critical stage" of culture development. It may also be borne in mind, in this connection, that there also exists just such a "critical stage" or "critical tube" in the serial transfers by which the bacteriophage is developed spontaneously from pure broth cultures, as demonstrated earlier by Hadley and Klimek. In these cases, the "critical tube" was most commonly one lying between the sixth and the fourteenth, depending on the conditions of the experiment. In the greater number of cases it was the eighth.

In the present experiment (table 2), the G forms appeared in greater numbers in the lithium broth series, and accompanied an exclusively S type population from the eighth to the twelfth tube. At this point, both S and G forms disappeared, giving a sterile plate on the twelfth test. Unfortunately, the lithium series was terminated at this point. At this time, moreover, we had not learned the advisability of maintaining the transfers regardless of the failure of growth to appear in the tubes.

In the plain broth series, which was continued, as also in the plain broth series of the experiment previously described (table 1), the G colonies present alone in the thirteenth and fourteenth plates gradually increased in size with further transfer until, by the twenty-eighth plate, the colony picture was that of a typical S type Shiga culture.

Dissociation Forced by Pancreatin.—Our notion of using pancreatin as a dissociating agent was based on observations by Hoder and Suzuki in 1926 that commercial pancreatin ("Pankreon") could serve as the source of bacteriophage. In 1927, Hadley and Klimek studied the influence of Squibb's and some other commercial pancreatins on Shiga and typhoid cultures with special reference to the generation of bacteriophage, and noted its strong dissociative effect on the culture of the substratum. It was concluded that the pancreatin employed did not serve as a source of bacteriophage, as Hoder and Suzuki believed, but that it served as an inciting agent which caused the bacteria to generate bacteriophage spontaneously as a result of dissociative behavior. In any case, the substance was a strong stimulus to microbic dissociation. It was primarily because of this fact that we were led to employ pancreatin in the experiments now to be reported.

The next three cases depicting the origin of the G type cultures concern the cultivation of the S or the R type Shiga culture in 5 per cent pancreatin (Squibb's) beef infusion broth and, for controls, in plain infusion broth. In one R series, the pancreatin broth had a reaction of 7.8, and its control had the same reaction (table 3). In the S series, one pancreatin broth had the reaction of 7.8 and its plain broth control the reaction 7.4 (table 4). Another pancreatin broth had the reaction 6.8 and its control the same reaction (table 5). Several points of special interest are to be found in the following details relating to the appearance or nonappearance of the G type cultures in the tubes of these series.[4]

Origin of Strain III: The first instance to be reported under this heading involves the serial transfer of an R type Shiga culture through pancreatin broth;[5] also a series of plain broth cultivations at the reaction 7.8. The original culture first employed was an R type produced by serial cultivation of a pure line S through 0.5 per cent lithium chloride broth. This R strain had the same origin as the culture that served as the point of departure for the production of strain I of the G form. The passages were made at twenty-four hour intervals and each tube, after incubation, was plated on plain infusion agar having a reaction of 7.4. Table 3 gives a record of this series and of the control series, extending from the first to the fiftieth tube in each series.

From the data presented in table 3 it appears that the first indication of dissociative action in both the pancreatin and the plain broth series occurred in the sixth tube of the series; and that, at this moment, the reaction was somewhat more pronounced in the pancreatin series. Here,

---

4. For the method of sterilizing the pancreatin broth tubes, see the section on control experiments on filtration and other tests, p. 76.

5. It may be noted here that, as shown in the earlier work of Hadley and Klimek, Squibb's pancreatin possessed a stronger dissociating action on the Shiga culture and on some other species than did two other commercial pancreatins. The reason for this is not known, nor the reason why pancreatin serves as a dissociating agent. We have used it in the present experiments because we had earlier recognized its strong influence in causing the generation of the bacteriophage in pure broth cultures of various organisms, and because this result was manifestly the outcome of the strong accompanying dissociative reactions. In other words, in those earlier tests, we produced the bacteriophage artificially by enforcing the dissociative reaction. In the present experiments, we are producing the G type culture by the same procedure. Owing to the complexity of the details to be followed, however, we have not attempted to study in detail the possible relation of the bacteriophage to the production of the G type cultures. Here, of course, we recognize the presence of a highly important problem to which we have not, as yet, been able to give undivided attention.

TABLE 3.—*The Origin of Strain III of the Shiga G Type Culture and Its Relation to the Accompanying Cyclostages*

	Strain III: 5 per Cent Pancreatin Broth						Broth without Pancreatin (No G Colonies)				
	Number of Colonies						Number of Colonies				
Tube*	S	SR	R	G	Remarks	Tube*	S	SR	R	G	Remarks
1	0	0	100†	0		1	0	0	100	0	
2	0	0	100	0		2	0	10	90	0	
3	0	0	100	0	Stable R	3	0	0	100	0	Stable R
4	0	0	100	0		4	0	0	100	0	
5	0	0	100	0		5	0	0	100	0	
6	11	44	45	0	First reaction to S (slight)	6	1	46	53	0	First reaction begins
7	6	32	62	0		7	1	33	66	0	
8	0	38	62	0		8	1	33	66	0	
9	0	10	90	0		9	45	8	47	0	
10	0	65	35	0		10	27	50	23	0	
11	0	0	0	0	Sterile	11	36	24	40	0	
12	0	0	0	+	First G	12	64	1	35	0	
13	0	20	80	?		13	78	19	3	0	Maximum S reaction
14	0	0	0	+	G only	14	6	40	54	0	
15	0	0	0	+		15	19	44	57	0	
16	0	0	0	+		16	0	90	10	0	
17	0	0	0	+		17	0	5	95	0	
18	0	0	0	+		18	0	4	96	0	
19	0	0	0	+		19	0	0	100	0	
20	0	0	0	+		20	0	0	100	0	
21	0	0	0	+		21	0	0	100	0	Second period of stable R
22	0	0	0	+		22	0	6	100	0	
23	0	0	0	+		23	0	0	100	0	
24	0	0	0	0	Sterile	24	0	0	100	0	
25	0	0	0	+		25	0	95	5	0	
26	0	0	0	+	2 G colonies	26	0	15	85	0	
27	0	0	0	+	2 G colonies	27	0	15	85	0	
28	Few	0	Few	+	Few G	28	0	50	50	0	
29	0	0	0	+		29	0	0	100	0	
30	0	0	0	+		30	0	50	50	0	
31	0	0	0	+		31	0	0	100	0	
32	0	0	0	+		32	0	0	100	0	
33	0	0	0	+		33	0	0	100	0	
34	0	0	0	+		34	1	2	97	0	Third period of stable R
35	0	0	0	0	Sterile	35	0	0	100	0	
36	·0	0	0	0	Sterile	36	0	0	100	0	
37	0	0	0	+		37	0	0	100	0	
38	0	0	0	0	Sterile	38	5	90	5	0	
39	0	0	0	+		39	10	90	0	0	
40	0	0	0	+		40	10	90	0	0	Partial reversion to S
41	0	0	0	+		41	7	0	93	0	
42	0	0	0	+		42	75	15	10	0	
43	0	0	0	+		43	0	80	20	0	
44	0	0	0	0	Sterile	44	0	4	96	0	
45	0	0	0	0	Sterile	45	0	5	95	0	
46	0	0	0	0	Sterile	46	0	0	100	0	
47	0	0	0	0	Sterile	47	0	10	90	0	
48	0	0	0	0	Sterile	48	0	0	100	0	Fourth period of stable R
49	0	0	0	0	Sterile	49	0	0	100	0	
50	0	0	0	0	Sterile	50	..	..	...	...	Discontinued

* Serial transfers, at twenty-four hour intervals, of an R type culture through pancreatin broth (strain III) of reaction 7.8, and its plain broth control (negative results) of the same reaction.

† Colony counts reported on basis of 100 colonies.

however, the reversal to R and RS quickly occurred; while, in the plain broth series, an intermediate stage was longer maintained. In the pancreatin series, the G type colonies first appeared in the plate from the twelfth serial tube, following a "sterile plate" (as was frequently observed in the experiments earlier described) arising from tube 11. The results from the thirteenth tube were uncertain; but in the following tubes the G culture persisted for a considerable time—in fact, to the forty-fourth tube in the series. In this long series of pancreatin tubes many of the plates were recorded as "sterile." Looking back on this and similar experiments, we doubt that either the broth tubes or the agar plates were actually sterile. They both undoubtedly contained the filtrable, but noncultivable, form of the organism. At that stage of the investigation, we did not have in hand the method later used to put this matter to the test. In a few cases, when these "sterile" plates were incubated for ninety-six hours or longer, a sprinkling of minute colonies was seen. It may also be noted that no G colonies were observed on the plates from the plain broth series. The culture undergoing transfer at last became stabilized in the R state. It is important to observe that this is the only instance in our records in which a strong dissociative action in the mother culture was not accompanied, at some moment, by the appearance of G colonies.

Origin of Strain IV: The results obtained in the experiment just set forth, involving the generation of G forms from an R type Shiga culture, suggested the experiment now to be described, in which the S type culture was employed. The S stock, on preliminary test, was found to yield 100 per cent S colonies. The culture mediums were the same as in the test detailed in table 3, except that the broth of the control (plain broth) series had a reaction of p_H 7.4 instead of 7.8 as in the pancreatin series. This was an inadvertence. The plain broth tubes were not therefore an exact control on the pancreatin series, but the results were nevertheless of interest. The serial transfers through both series were made at twenty-four hour intervals, and the platings on plain infusion agar were carried out as before. Table 4 gives the results on these two series through the twenty-seventh tubes. The record of the first appearance of the G type colonies is presented separately.

From the results presented in table 4 it appears that the serial transfer of the pure line Shiga S type culture through twelve tubes of 5 per cent pancreatin broth yielded the G forms; that these subsequently disappeared and were absent in the next six tubes, but that they reappeared in the eighteenth tube and then persisted through the rest of the series nearly to its termination at the twenty-seventh. The first and second advents of the G colonies, as was so often the case, were preceded by "sterile plates." In the plain broth series, on the other hand, the S type culture showed marked stability; no sign of dissociative action was observed, and no G colonies were encountered, throughout

TABLE 4.—*The Origin of Strain IV of the Shiga G Type Culture and Its Relation to the Accompanying Cyclostages* *

Strain IV: 5 per Cent Pancreatin Broth					Broth without Pancreatin (No G Colonies)					
Number of Colonies						Number of Colonies				
Tube*	S	SR	R	G	Remarks	Tube*	S	SR	R	G	Remarks
1	0†	100	0	0		1	100	0	0	0	
2	30	70	0	0		2	100	0	0	0	
3	100	0	0	0							
4	100	0	0	0							
5	100	0	0	0							
6	5	10	85	0	Maximum R						
7	2	67	31	0		Throughout 27 serial transfers, the results were the same as indicated above for tubes 1 and 2. No G colonies were observed in any plate, with the possible exception of the nineteenth, which was doubtful
8	0	100	0	0		
9	2	24	74	0		
10	100	0	0	0		
11	0	0	0	0	Sterile	
12	3	36	61	+	First G	
13	100	0	0	0							
14	100	0	0	0							
15	100	0	0	0							
16	100	0	0	?							
17	100	0	0	0							
18	0	0	0	0	Sterile¶						
19	100	0	0	+	Second G						
20	100	0	0	+							
21	100	0	0	+							
22	100	0	0	+							
23	100	0	0	+	Only few G						
24	100	0	0	+							
25	100	0	0	+							
26	100	0	0	+							
27	100	0	0	0							

* Serial transfer, at twenty-four hour intervals, of an S type Shiga culture through 5 per cent pancreatin broth of reaction 7.8; also through plain infusion broth of reaction 7.4 (negative results).

† Although this test was started with a pure-line and apparently well stabilized S type culture, the residence for twenty-four hours in the pancreatin broth enforced a transformation to SR. The culture returned to S, however, after two further transplants and then entered the S to R transformation.

¶ After seven days' incubation, this plate showed a small number of G colonies.

the period of the test, except possibly in the nineteenth tube. To what extent this result was due to the reaction of the control medium (p_H 7.4) cannot be stated; other tests suggested that this could not be the cause of the difference. We know, however, that an S type culture shows greater stability in the more acid mediums, just as the R type culture manifests greater stability in the more alkaline mediums.

Origin of Strain V: After having produced the G type culture by serial transfer of the S type culture through pancreatin broth of reaction p_H 7.8, we made a further attempt to accomplish the same result by the use of broth of p_H 6.8. In this case, the plain broth control had the same reaction. In this experiment it was suspected that the stabilizing influence of the acid broth on the S type culture might counteract the positive dissociating influence of the pancreatin, and in this manner prevent the generation of the G type culture. The same purified S type culture was used for this test, and all other points of

TABLE 5.—*The Origin of Strain V of the G Type Shiga Culture and Its Relation to the Accompanying Cyclostages*

Strain V: 5 per Cent Pancreatin Broth					Broth without Pancreatin (No G Colonies)				
Number of Colonies					Number of Colonies				
Tube* S	SR	R	G	Remarks	Tube* S	SR	R	G	Remarks
1 0	100†	0	0		1 100	0	0	0	
2 100	0	0	0						
3 100	0	0	0						
4 100	0	0	0					Throughout serial transfers 1 to 17 in
5 100	0	0	0					this control series, no dissociative action
6 98	0	2	0	S to R dissociation				and no G forms were observed. All plates gave 100 per cent S colonies
7 100	0	0	+	First G					
8 0	0	0	0	Sterile					
9 100	0	0	+						
10 100	0	0	+	Few G					
11 100	0	0	+	500 G					
12 100	0	0	+						
13 100	0	0	+	Few G					
14 100	0	0	+	G disappearing					
15 100	0	0	+						
16 100	0	0	+						
17 100	0	0	+	10 to 25 G					
18 100	0	0	0	G disappears					
19 100	0	0	0						
20 100	0	0	0						
21 100	0	0	0						
22 100	0	0	0						
23 100	0	0	0						
24 100	0	0	0	Termination					

* Serial transfer, at twenty-four hour intervals, of an S type Shiga culture through 5 per cent pancreatin broth of reaction 6.8; also through plain infusion broth of the same reaction (negative results).
† See footnote to table 4.

technic were the same as in the experiments previously reported. The results of these tests will be found somewhat different from those given in table 4.

From the results presented in table 5 it is seen that the G type culture first appeared in the pancreatin broth series in the seventh tube of transfer, but failed to appear at any time in the plain broth series. In the latter there was never any indication of dissociative action; and even in the pancreatin series the evidence of dissociative behavior was slight. In the pancreatin series, the first appearance of the G colonies in large number was immediately followed by the "sterile plate" (plate 8). After the appearance of the G colonies in large number in plate 9, the number was gradually reduced, and the last seven plates in this series showed none. The G culture did not become stabilized.

Another point of interest in this experiment must be noted. In the pancreatin broth series just reported, we found the G type culture appearing accompanied by only mild evidence of dissociative activity in the culture mass (tube 6). At the same time, this was the only occasion on which any dissociative action was observed in the entire series of twenty-four tubes. It can scarcely be doubted that, in this case, the agency that held in check the dissociative action of the mother culture was the acid reaction of the medium. Apparently the G forms may arise in cultures in small numbers despite the absence of strong dissociative action. But it may also be noted that the total number of G colonies was small, and that they did not persist in the culture, which remained stabilized in the S type. We are inclined to believe that this instance does not constitute an objection to our view that the generation of the G forms is a direct result of dissociative action. It merely serves to raise the question as to the influence of the reaction of the medium on dissociative activity and, pari passu, on the generation of the G type cultures. Into this phase of the problem we have hardly penetrated.

Origin of Strains XI and XII: These were obtained from apparently sterile bacteriophage filtrates. The method was the same for both strains, and only one instance will be presented.

A fresh sample of bacteriophage was built up from a laboratory Shiga stock (S type). The titer was approximately 10^{-9}. The broth was filtered through a Berkefeld N candle under 200 mm. of negative pressure, and the filtrate dispensed into tubes. The sterility was tested at once by seeding three purple agar plates with 2, 5 and 10 drops, respectively. These plates after incubation at 37 C. for forty-eight hours were apparently sterile. At this time, the lytic filtrate showed no indication of clouding. After incubation for ninety-six hours, it was placed at room temperature until the eleventh day. It was then subjected to the test involving the serial plate method. Agar plates were spread with 10 drops of the filtrate and incubated. Examination at the end of twenty-four hours showed no sign of growth. The plate was washed over with sterile broth, and the "washings" were transferred to a second plate. The same process was carried out three times. On the fourth serial plate appeared a fine sprinkling of thousands of minute colonies. From these the G strain was established. After further aging, the original samples of filtrate showed no sign of culture development. They were obviously sterile, but, as shown, the appearance was misleading.

We are of the opinion that the majority of lytic filtrates contain living micro-organisms the presence of which can be demonstrated if the proper methods are employed. The ease of such demonstration is increased as the filtrates become older.

Dissociation Produced by Aging.—Origin of Strain XIV: The instances reported in the foregoing pages involved the production of the G forms in cultures of the S type or of the R type Shiga undergoing serial transfer in mediums of various constitutions; it has been seen that the G elements appeared in young cultures. The question now naturally arises whether the G type culture might not appear in a single tube left to age for a sufficient time and examined at intervals by the usual plating method. This is suggested by the circumstance that other dissociations, especially those involving S to R transformations, often occur in aging tubes. Several experiments of this sort were performed, of which the one to be described is an example.[6]

A tube of plain infusion broth (p_H 7.8) was inoculated with one loop of an S type culture. This culture, as revealed by plating immediately before the test, was 100 per cent S. After a brief incubation at 37 C., the tube was allowed to stand at room temperature for fifty-eight days. Examinations were made at intervals by plating on infusion agar, in order to ascertain the differential colony counts. The results are presented in table 6. The colonies of intermediate type (between R and S) are given under the headings Sr and sR, the former more resembling S colonies and the latter more resembling R colonies in size, shape, color and contour. The numbers in the various columns refer to percentages, except in the column headed G. Here the number indicates the approximate number of G colonies actually observed.

From the data presented in table 6 it is apparent that the first evidence of dissociative variation in the original S type culture was observed on the plate made on the seventh day. At this time there was a distinct transformation toward the R colony form; the S colonies had already disappeared and 3 per cent of the colonies were pure R. On the plates of the ninth and eleventh days, however, there was evidence of retransformation to the S type (60 per cent), and 40 per cent of the colonies were intermediates (Sr). On the eleventh day, the G colonies first made their appearance on the plates, and they may have been present on the tenth day. The G forms persisted for a time and

---

6. In reality the test here described was a control test on one performed to reveal the influence of a G culture filtrate on the development of an S type culture. The results of the primary test were not of sufficient interest to justify further mention at this time. The control, however, had points of special interest.

were again detected on the sixteenth and on the twenty-second days. After this they disappeared, and the culture returned to 100 per cent S type. In this state it remained during the remainder of the experiment, which was concluded on the fifty-eighth day.

In these observations one may again see some of the characteristic points in culture behavior already delineated for the cultures undergoing serial transfer in plain broth or in modified culture mediums. First, the G forms did not appear until there had been observed a beginning of dissociation from the original S type culture. Second, there was a moment when the S forms had completely disappeared from the culture preceding the appearance of the G forms. Third, the G colonies arose

TABLE 6.—*The Origin of Strain XIV of the G Type Cultures (B. Dysenteriae Shiga) in a Plain Infusion Broth Culture of p_H 7.8, Standing at Room Temperature for Fifty-Eight Days*

Day of Observa- tion	Type of Growth (Colony Form) Observed on the Agar Plates					Remarks
	% S	% Sr	% sR	% R	G*	
1	100	0	0	0	0	S stable
2	100	0	0	0	0	S stable
3	100	0	0	0	0	S stable
4	100	0	0	0	0	S stable
7	0	0	97	3	0	First dissociation
9	60	40	0	0	0	
11	60	40	0	0	125	First G
16	100	0	0	0	100	G remains
22	100	0	0	0	100	Last G
26	100	0	0	0	0	G disappears
29	100	0	0	0	0	
58	100	0	0	0	0	S stable

* Approximate number of colonies per plate.

on the plates at about the same time (from the tenth to the eleventh day of the experiment) in the life of the culture. Fourth, after a brief sojourn, they disappeared, being replaced by the S type cells.

From these results, as also from certain other tests of a similar nature, mentioned briefly on an earlier page, we may conclude that serial transfer is not necessary for the generation of the G elements in an S type culture; they may arise as well in single broth cultures if these are allowed to age, and may be detected if the examinations are made at a favorable moment in the development of the culture.

Origin of G Strain of B. Coli.—This case is included, although it does not concern the Shiga bacillus, because it reveals a different manner of origin of the G forms, and because it offers the suggestion that these elements arise directly from the cells of the R type culture rather than directly from those of the S type.

An S type *coli* was being transferred at twenty-four hour intervals through 0.35 per cent lithium chloride broth of p_H 7.8 for the purpose of producing G colonies. Platings from each tube in the series were made at the end of the period of growth. The trend of the differential colony count is shown in table 7.

At plate 11 in the series, the procedure was modified and one of the R colonies appearing on the eleventh plate was selected as the inoculum into the twelfth tube of broth. It was believed that this procedure might hasten the process of transformation of the bacterial population as a whole. This actually occurred, as is indicated by the appearance of 20 per cent R colonies on plate 12. The R colonies were not perfect, however. The central areas of most of the colonies showed some

TABLE 7.—*Differential Colony Counts in Experiment to Produce G Forms of B. Coli*

Coli Form	Counts for Each Tube and Plate in Series *											
	1	2	3	4	5	6	7	8	9	10	11	12†
S	100	100	85	80	80	50	35	25	25	25	15	0
SR	0	0	15	20	20	50	85	75	75	75	80	80
R	0	0	0	0	0	0	0	0	0	0	5	20 →G§
G	0	0	0	0	0	0	0	0	0	0	0	0

* The numbers in the table represent approximately percentages of the different colony types appearing on the plates.
† Inoculated from an R colony on plate 11.
§ From an R colony.

S structure, while the free edge, where active growth was taking place, was advancing rapidly into the pure R state. A plating was therefore made from the edge of an imperfect R colony, which was made up largely of filamentous and swollen rod forms, as shown in stained preparations. The edge was touched with the needle and the inoculum diluted in a tube of plain broth. The dilution was then spread on plain agar plates. After twenty-four hours there arose several thousand colonies ranging in size from 0.3 mm. down to the limit of vision ($10 \times$ hand lens). The average size was about 0.2 mm. In view of these results, the broth dilution tube mentioned was filtered through a Berkefeld N candle under 200 mm. of negative pressure. From the filtrate the G forms were obtained by the usual serial plate method.

Origin of G Forms in Secondary Colonies of S Type Mother Culture.—The following facts concern the origin of the G type culture from a somewhat different point of view, namely, as secondary G colonies within mother colonies of the S type. This instance concerns, not the dysentery bacillus, but *B. typhosus,* studied by Miss Bailey in our laboratory. The data are included here because, as will appear later, they have an important bearing on the possibility that the G form might be regarded as a foreign bacterial species living in close association with the principal culture submitted to study.

A pure line S type culture of *B. typhosus* was being passed at twenty-four hour intervals through a plain infusion broth medium of reaction 7.8. Plates streaked from the first and second broth tubes in the series showed only normal S type colonies and no G forms. The plate from the third serial broth tube revealed, at four days, only S type colonies and no free G forms. When, however, this plate was examined on the tenth day, the majority of the S colonies were found to contain minute, secondary colonies, some central, others peripheral. These showed, under a no. 3 objective and by transmitted light, as dark granules, and by reflected light, as minute papillae measuring not more than from 0.05 to 0.1 mm. in diameter, rising slightly above the surface of the mother colonies. The majority of the S colonies showed these papillae—some only one or two, others a dozen or more. The fourth plate in the series, when aged from 6 to 10 days, manifested not only the papillary mother colonies, but also, and for the first time, free G type colonies lying among the S forms. Plates streaked from the fifth and sixth tubes in the series showed decreasing numbers of G colonies, and the papillary character of the S type colonies was lost.

When one of the papillary colonies was suspended in sterile broth and a dilution of this streaked on an agar plate, there appeared, after about twenty-four hours' incubation, colonies of all sizes from the "normal" S (about 2 mm.) to some colonies from 0.2 to 0.3 mm. in diameter. Some of the smallest were picked to broth, incubated and the broth filtered. Streaking several drops of the fresh filtrate on litmus lactose agar plates revealed no visible growth after forty-eight hours' incubation. Further plate-to-plate transfer of the plate washings, however, eventually caused the generation of typical G colonies, identical with those obtained from the stock culture on several occasions by other methods.

It seems to us that these observations made by Miss Bailey strongly suggest that the G type cells may arise from the S type cells, either before plating is carried out (in which case they register on the agar plate as discrete G colonies) or after the plating has been accomplished (in which case they appear as secondary colonies within the S colonies). In other words, the aged S type cells, at a certain moment in development, give birth to the G forms. If the S cells attain this favorable stage shortly after plating has occurred, the G colonies appear as secondaries within the S. If, on the other hand, the S cells attain this stage, and therefore generate the G elements, before the plating occurs, the G forms are already free and can register in their characteristic colony form, not within the mother colonies, but independently on the surface of the medium. It is thus strongly suggested that the first generation of the G elements is dependent on the attainment by the S type cells, of a certain stage of development. Whether this stage is reached in the broth tube or in the colony mass has no influence on the final outcome. The dissociating culture, in other words, must pursue the clearcut path of its specific cyclogeny; and, so far as the G forms are concerned, this phase of culture existence may be attained either in liquid or on solid mediums. We shall have occasion to observe later, however, that there exist associated with these G forms living corpuscles of some sort, on the further development of which the nature of the medium exerts a definite influence.

These details relating to secondary colony formation by the G forms within the S colonies have dealt with the typhoid bacillus and not with the Shiga organism. As we have stated, however, we have reason to believe that the train of events that occurs in one species is duplicated in the other; and that it has been due merely to mischance and to our failure to keep certain of the Shiga plates under observation for a sufficient time that we have missed the important picture which Miss Bailey was so fortunate as to discover.

CYTOLOGIC ASPECTS OF THE PROBLEM OF ORIGIN

In connection with the origin of the G type cultures there remains the question of their origin cytologically—that is, the manner of actual emergence or generation from the parent cells of the culture; for it must be admitted that they arise in some way from the well known culture elements. On the manner of origin in all cases we have no certain knowledge. There do exist, however, certain possibilities that cover all cases; and these are based on observations that we have made on stained preparations and dark field mounts prepared at various moments in the development of the cultures which were under the influence of lithium chloride, pancreatin or some other inciting agent.

By various methods it was found easily possible to detect in cultures about to generate the G forms minute granular bodies, varying considerably in size, shape and staining characteristics, but yielding the impression that they were living elements of some sort. We freely admit, however, that it was often a difficult matter to differentiate between the granules that were undoubtedly artefacts (perhaps representing colloidal substances in the medium) and the bodies that we believe were living elements. If it had not been possible to filter them and to recover them from the filtrates, we should still have slight evidence that some of them were other than colloidal particles. Further, we are of the opinion that there existed in these cultures, amid the visible living granules, other granular elements lying beyond the range of microscopic vision. Whether it is the invisible granules only that are able to pass the filter candles, is still a question.

Regarding the manner of origin of these viable granules from previously existing cells, several possibilities may be mentioned. In the first place, we have observed in cultures, both before and after the elaboration of the G forms, many common rod forms that appeared to be undergoing granulation. The number of granules per cell varied from one to three or four. The size was fairly constant (from 0.3 to 1 micron) and compared favorably with similar bodies observed outside the cells. Some of those observed outside were smaller than any observed inside the cells. Actual liberation of the granules from the cells was not observed.

In the second place, we have noted in similar cultures rod forms of normal appearance that were budding off minute granules either laterally or terminally. In some cases, the first bud formed appeared to be giving off a secondary and somewhat smaller bud. In some cases, the granules were detected lying close to the wall of the cell, causing slight swellings or so-called "hernias" of the cell wall. These granules appeared analogous to those seen by us in several other bacterial species. They were often very distinct in cultures of *B. subtilis* and *B. anthracis;* also in cultures of *B. typhosus*. Indeed, as Löhnis (1921) pointed out, similar structures were first pictured by Koch in 1877 in one of the earliest photomicrographs (original fig. XVI, 5) of the anthrax bacillus. Koch termed these bodies "seitliche Sporen." We have not been able to ascertain, however, that he ever made any further reference to them, and it may be believed that he eventually became converted to the view, prevalent at the time and maintained since by the majority of bacteriologists, that these "spores" represented merely the extrusion of incidental materials (such as degenerate protein, wax droplets or fat droplets) from the cell. It is of interest, however, that Koch still maintained his views regarding the "spores" (granules) in *B. tuberculosis*.

In the third place, we have seen, especially in cultures under the influence of lithium chloride, bodies that we believe are analogous to the Pettenkofer bodies of Kühn, also to the so-called "Ferransche Körperchen" and related bodies often described in the bacteriologic literature between 1878 and 1890, but seldom mentioned since, except under the name of artefacts and involution or degeneration forms.

These bodies appeared as enlarged cells, usually round or oval; they have sometimes been referred to as "balloon bodies." The diameter varied from 2 to 7 microns, the average size being about 4 microns. When examined by dark field methods or in preparations stained with simple stains, they gave no features of special interest. When stained by the Giemsa method, they were often seen to contain from one or two to a dozen nucleus-like granules, staining reddish purple. We have never observed the liberation of these granules from the "sac," although several other investigators have reported success in such attempts— among them, Kühn. We have, however, noted bodies that appeared flattened and "empty," as if the contents had been liberated. We have also seen similar bodies in cultures of a streptococcus derived from the virus of encephalitis. Kühn presented excellent photomicrographs of these elements. It should be borne in mind, however, that his explanation and interpretation of these pictures is quite different from ours, for he regarded them as evidence of parasitism of the bacterial cell by a myxomycete, the spores of which eventually escaped from the "balloon bodies."

Finally, we have observed in the filaments of the Shiga cultures of the R type, and in many other bacterial species, the small, deeply staining granules spoken of in earlier days of bacteriology as "gemmules" and often described and pictured for *Leptothrix*. We have not observed the actual liberation of these granules from the parent sheath, but we have no doubt that it occurs. We have often seen the empty sheaths, pale-staining and destitute of visible contents. Outside the sheaths similar granules could be observed in considerable number. These are no doubt similar to the granules of *B. welchii* observed in the R filaments of this organism by Roe in this laboratory; also to the granules in the R filaments of *B. acidophilus* recently reported, with photomicrographs, by Faith Hadley (1930, b). There can no longer be any doubt but that these granules constitute elements of reproductive significance in the filaments of the R cyclostage.

The clearest evidence of at least one mode of origin of the G forms and the G colonies was obtained from an R colony of *B. coli*.

This was still an imperfect R and had been obtained from a pure S colony as a result of growth in 0.35 per cent lithium chloride broth. A plating from the twelfth tube in the series had revealed fair R colonies, although the inner portions still contained a considerable number of S elements. Up to this point no G forms had appeared. The margins of the colonies, after five or six days of growth, became progressively more R-like, giving the typical "cut-glass" appearance common to the R forms of many species. When this plate was 8 days old, the microscopic picture of the edge showed large spherical bodies, swollen rods and long filaments with peculiar swellings. Many of the filaments contained small granules. An attempt was made by further plating from the edge of this colony to secure a more perfect R type.

Accordingly, a needle was touched to the edge of the colony and the inoculum diluted in plain broth of p_H 7.8. Without delay, this was spread on a plain agar plate of p_H 7.4, and the plate incubated for twenty-four hours. On examination, this plate revealed, not the expected R colonies, but only G colonies and a very few GS intermediates. No S, SR or R colonies appeared.

When cover glass impression films were made from the G colonies, they revealed, curiously enough, not typical G elements, but chiefly swollen rods and filaments. These, however, were unlike those of the original R colony. They scarcely took the stain and appeared like dead cells. Occasionally the rods and filaments contained minute, deeply staining granules. Similar granular bodies and cocci were scattered between the rods and filaments.

From these results we tentatively conclude that, when the swollen rods and filaments were taken from the edge of the R colony, diluted in broth and spread on the plate, they began to develop as rods and filaments, but that this process was abruptly terminated by the liberation of the granules. The large, pale-staining forms, we therefore conclude, were actually dead cells, their function in the production and liberation of the granular bodies having been fulfilled. In other words, very shortly after their deposition on the agar surface, they reached the moment of reproductive maturity which, at the same moment, marked the culmination of their vegetative existence and the consequent death of the old cells. This exposition, whether the interpretation is correct or incorrect, clearly illustrates the extent to which the discovery of some of these peculiar phenomena of culture development is due to chance.

As has been pointed out earlier, a *coli* G culture was derived from the filtration of a broth tube seeded from the plate containing the G colonies; also from a filtrate of the dilution tube from which the G colonies were first obtained.

In concluding these paragraphs on the cytologic origin of the G forms and the associated filtrable elements, it appears that there is some reason to believe that the generation of these bodies stands in relation to the "granules" of bacterial cells; and we believe that many of these possess a reproductive function. That these granules, after liberation from the cells, undergo further division into still smaller bodies seems probable, since we saw good evidence of this fact in the case of the lateral gonidia of the anthrax bacillus. The same phenomenon has been reported by others for the cholera vibrio. Enderlein also described further division of the gonidial bodies. Since many of the gonidial forms, when first liberated, possess a considerable size, it seems possible that it is only the products of further division, the microgonidia, that are able to pass candles ordinarily employed for filtration studies. With the foregoing statements of our observations on this aspect of the subject we prefer to let the matter rest for the present.

SUMMARY OF SECTION AND CONCLUSIONS

The foregoing exposition is perhaps sufficient to indicate the experimental conditions under which the G type cultures first arose. These experiments have been presented merely as examples of many others that were conducted by more or less similar methods and that invariably yielded the same order of results. We hope that it will be noticed particularly that the G type cultures described were not produced by resort to filtration, as has usually been the case in work of this character and particularly in the case of the filtrable form of the tubercle bacillus. The G forms of the Shiga bacillus were obtained by setting into operation the dissociative reaction in the mother culture. The G forms, moreover, did not appear as the culmination of any gradual, grading-down process in the colonies of the original culture nor by long-continued artificial selection of smaller and smaller colonies. The G colonies always appeared abruptly, as a discontinuous, rather than continuous, variation. In this respect, they simulate mutations; or perhaps sudden invasions of the mother culture by the cells of a contaminating species. And as such they will doubtless be regarded by many of our readers, particularly by those to whom dissociative reactions are still much of a mystery. For the present we shall not attempt to demonstrate that this is not the case. The consideration of this possibility will receive attention later. We shall therefore speak of them merely as the G type cultures, observed on numerous occasions in association with purified and "normal" S and R cultures of the Shiga dysentery bacillus (and several other bacterial species) in the state of active dissociation.

In the lithium chloride broth series (two experiments reported), the G colonies appeared in both the main series and the control series. In the pancreatin broth series (three experiments reported), the G colonies arose at some time in all three of the main tests, but not in the broth controls. The failure in the latter we are inclined to attribute to the low reaction of the medium. In several tests, the G forms finally supplanted all other forms of culture. In other cases, they continued to exist in association with the S forms for a considerable time, then disappeared. In four experiments, the first appearance of the G forms in numbers followed closely, or was accompanied by, the disappearance of all other forms of culture. The S forms, previously present in abundance, seemed to undergo lysis in the generation of, or transformation into, the G elements, or into some culture state that preceded them.

The regularity with which the G forms were obtained under the influence of pancreatin broth, while lacking in the plain broth controls, might seem to throw some suspicion on the pancreatin as the source of the new culture type. In answer to this objection it may be said that the presumably sterile pancreatin broth was given rigorous tests, some of which will be presented later in the section on controls. Moreover, it is apparent that the G culture made its appearance just as regularly in other series of tests in which the pancreatin was replaced by sterile solutions of lithium chloride, and even in other cases in which no additions whatever were made to the sterile broth mediums. In other words, we cannot associate the appearance of the G type colonies with one kind of medium more than with another, although the presence of lithium chloride or pancreatin, especially in conjunction with a high p_H reaction (7.8), seemed to intensify the dissociative reaction. That even these substances, however, are not necessary is demonstrated by the number of cases (two of five) in which the G forms arose in unmodified mediums possessing sufficient alkalinity.[7]

The time of appearance of the G forms in the mother culture is also a point of considerable interest. Whatever the nature of these G strains may be, we can note that they are intimately associated with the Shiga cultures, and that they may be derived from growths that have been started with either the S or the R forms. We can observe, moreover, how hopeless it is to anticipate the presence of this new culture type in every culture, or even in a single culture at all times in its development. The G forms make their first appearance at a certain moment, which, in our experience, is most likely to be between the sixth and twelfth serial tubes in a series undergoing twenty-four hour transfers. The majority of the G cultures have first appeared in the tenth, eleventh or twelfth tube. Their advent may take place in mediums that would be regarded as favorable for the development of the "normal" Shiga culture, although their appearance seems to be made more certain when there

7. This point is also supported by observations on the production of the G forms in other bacterial species.

is brought to bear on the young culture certain influences that are known to favor the dissociative reaction, as, for example, alkaline mediums, the presence of lithium chloride or of pancreatin in the broth, mediums containing peritoneal fluid, or aging.

We have seen, moreover, that the duration of these new forms in a culture undergoing serial cultivation in broth mediums, or even in standing tubes, may be very brief, at least so far as their revelation on the agar plate is concerned. Either the S or the R form of culture may gain the supremacy, and the G disappears. On the other hand, we have noted certain instances in which the G forms alone maintained their position, to the complete exclusion of all other culture elements, and did so over a considerable period of time. Moreover, there were cases in which the G culture, after maintaining its existence through numerous passages, gradually transformed in its colony form and cultural features to the original mother type. And finally, as has been shown, it was possible for the G forms to appear, to remain for several passages, to disappear for many more, and then to return again to their position among the other culture elements. Regarding the conditions that determine or facilitate their appearance, their residence, their disappearance or their reappearance, we know little. In this connection, however, much interest attaches to the circumstance that in so many instances the first appearance of these forms was immediately preceded, or sometimes accompanied, in the transmission series, by a "sterile plate." The real significance of this is not recognized at the present moment, but it appears as if the great mass of preexisting culture forms, or a large part of them, become expended in some way in the process of generating the new elements of the G type culture, or in the generation of some still unrecognized, invisible elements that precede them on the agar plates. It is probable that in some cases the predecessors of the G elements arise from the R forms.

Regarding the cytologic origin of the G forms from the previously existing cells of the culture, our knowledge may be summarized by the statement that we have observed in the mother cultures the origin of small granular bodies arising in various ways from the single cells and filaments. Some of these may represent the filtrable elements. On the other hand, it is possible that the actually filtrable elements are granules which lie below the range of microscopic vision. These may be formed directly by the larger cells or filaments, but they probably result from the further division of the granules first produced by buddings from the rods or as "gemmules" within the longer filaments.

These are, then, the original sources or methods of production of some of the G type cultures the further characteristics of which will be considered in the following pages. And it should be borne in mind that what has been stated regarding the Shiga cultures also holds true, in a large degree, for many other bacterial species as demonstrated by ourselves and other workers in our laboratory group.

In conclusion, we may summarize the methods by which we have been able to cause the generation of the G type cultures from either the S or the R forms, as follows: (1) by serial passage through plain infusion broth of reaction p_H 7.5 to 7.8; (2) by serial passage through lithium chloride broth; (3) by serial passage through pancreatin broth; (4) by the use of peritoneal fluid; (5) by the use of bacteriophage; (6) by aging on agar slants, or in tubes of alkaline broth; (7) by selection of secondary G colonies in S type mother colonies, and (8) by cultivating from the edge of an old R type colony at the moment when the cells reach reproductive maturity.

While the most constant results were obtained by the use of lithium chloride and pancreatin, we believe that many other chemical substances, organic or inorganic, could accomplish the same results. Indeed, Plantureux (1930) showed that the chlorides of calcium, strontium and barium, added to alkaline broth mediums, possess strong dissociative power. Moreover, Oesterle and Stahl recently demonstrated the influence of sunlight, of ultraviolet light, of high concentrations of sodium chloride and of soda, and of minute amounts of mercuric chloride and of chloramine-Heyden in producing the filtrable forms of *B. mycoides*.

None of these statements is meant to imply, however, that we believe that the generation of the G forms is dependent on the presence of unusual or "injurious" substances in the medium. For the present, we prefer the view that the generation of the G elements is a part of the normal reproductive activity of the culture at certain periods of its existence. That this normal reaction may be hastened or intensified by modifying the environment of the growing cells, either chemically or physically, appears to be shown by the experiments reported in this section.

APPEARANCE AND BEHAVIOR OF G TYPE CULTURES ON AGAR AND IN BROTH MEDIUMS

It has perhaps been made clear in the foregoing section that the G type of colonies, on their first appearance on plates, manifest characteristics sufficiently clear to make certain their differentiation from other known colony forms, such as the common S, SR, R and other intermediates. We now wish to present the appearance of the colony in greater detail and, in addition, to describe certain peculiarities of growth which, so far as we are aware, are not encountered in any other type of culture, and have not been reported in detail among the common forms of growth of any bacterial species.

APPEARANCE OF G GROWTH ON AGAR PLATES

In all the cases mentioned in the preceding section, the first appearance of the G colonies was on beef infusion agar plates seeded from broth cultures undergoing the dissociative reaction. For purposes of description these colonies may be divided into four groups: (1) those visible to the unaided eye; (2) those not visible to the unaided eye, but visible with the aid of a hand lens magnifying from 3.5 to 10 diameters; (3) those not distinctly visible with a hand lens, but visible with the no. 3 objective, and (4) those not visible with the no. 3, but visible with the no. 7, objective. The smallest colonies observed at the end of forty-eight hours' growth at 37 C. had a diameter of from 0.004 to 0.006 mm. The largest in a freshly isolated G strain had a diameter of about 0.2 mm. after four days' growth. The average size of the S type colonies grown on the same medium for the same length of time was 2 mm.; that of the R colonies, under the same conditions, from about 4 to 5 mm. The smallest of the G colonies were studied to advantage by dropping a cover glass over the agar surface and examining with the no. 3 or the no. 7 objective. Often the cover glasses were removed and the impression films stained by Giemsa. Many plates containing multitudes of these minute colonies would have been regarded as sterile in the course of ordinary observation of plates.

In all cases, except two, the G colonies, at their first appearance and when observable under a hand lens, were round, regular and smooth. By transmitted light, they were bluish or translucent; rarely, somewhat whitish. The whitish appearance increased as the G forms, by very gradual stages, approached the S type. This change was accompanied by a gradual increase in colony size. In two instances, the G colonies were irregular in shape and revealed broken or serrated margins. One of these "rough type G colonies" reverted, after many passages, to a typical R type Shiga culture.

Since it was found to be a matter of difficulty to subculture a colony measuring from 0.05 to 0.2 mm. with a platinum needle, much of this work was done with sterile, fine, glass threads freshly drawn out from pieces of 4 mm. rod stock. To prevent vibration in handling, the needle portion was not more than 5 cm. in length and reasonably stiff. The undrawn rod constituted the handle and was of any convenient length. The selections of colonies were made only after examination of the field under a no. 3 objective. The subculturing itself had to be performed under a hand lens or under the no. 3, depending on the size of the colony. A binocular microscope equipped with arm rests is convenient. After the selected colony had been touched with the fine tip of the glass needle, it was touched off to broth, usually tubed in 2 cc. amounts, and incubated for varying times, usually from forty-eight to seventy-two hours. Such inoculations to broth yielded evidence of growth in due time in about 75 per cent of cases. Mass cultivation from small groups of colonies was successful in the larger percentage of attempts, but attempts to cultivate from a single colony directly to agar by the "dry" method commonly failed.

It will now be understood by the reader that such cultivations in broth as have been described, and the detailed appearance of which will be described later, represented our "stock strains" of the G type cultures. It was these broth cultures, sometimes arising from single colonies directly, sometimes from Berkefeld filtrates as will be pointed out later, that were made use of for the studies here reported. In the following sections, we will present some of the characteristics of these strains and especially those dealing with their morphology and manner of growth in broth and on agar.

The attempted propagation of these G colonies gave results possessing special interest, depending on whether they were freshly isolated from filtrates (see section on filtration) or obtained from the stock G cultures aged from 6 to 50 days.

When the Berkefeld filtrate of a G type culture was allowed to stand at room temperature for from six to twelve days, it often happened that fine granulations gathered on the wall of the tube and gradually sedimented, yielding eventually a viscogranular sediment at the bottom of the tube. The tubes seldom became cloudy short of several months, and even then the appearance was more of an opalescence or faint haziness. Cultivations attempted from the sediment during the first five or six days usually failed to reveal on plates any sign of colony structure on the first or second serial plates. The third or fourth serial plates, however, usually showed distinct G colonies in fine clusters on the surface. After from six to twelve days of standing, the inoculation of the filtrate sediment to agar plates more commonly gave G colonies on the first or second plates of the series. After the filtrate had stood for several weeks to two years or more, inoculation would frequently yield colonies on the first plate seeded, although from six to ten days were often required for incubation.

If some of the colonies that appeared first were streaked on a second plate, no visible growth was likely to appear within from twenty-four to forty-eight hours. If, on the other hand, the colonies from the first plate were planted in broth and the broth growth was incubated and then streaked on the plate, the G colonies appeared.

When these colonies, in turn, were streaked on agar, no sign of growth usually appeared. Yet, if they were placed in broth and the broth growth was streaked on agar, the G colonies developed. This phenomenon was observed many times and was not limited to the Shiga cultures. In only a few cases did we secure evidence that freshly isolated G colonies could perpetuate their colony form of growth by agar to agar transfers. It seemed necessary that a brief sojourn in broth should intervene. An explanation of this phenomenon is not at present available. That it is not an accidental occurrence in the G type cultures, however, is indicated by its peculiar periodicity; also by the fact that a somewhat similar train of circumstances has been fully established for the G type of B. typhosus and several other bacterial species studied either by ourselves or by others in this laboratory.[8]

APPEARANCE OF G GROWTH IN BROTH

This varied with the source of the G strain and with its stage of development.[9] When the G cultures first appeared on agar plates,

8. Other data bearing on the appearance of the G type cultures and certain peculiarities of their behavior will be presented in the section on filtration experiments, p. 49; also in the section on viability, p. 87.

9. The meaning of the phrase, "stage of development," considered more fully on a later page, may be clearer to the reader if we state that we have come to regard the G type culture as an intermediate stage of culture development lying between certain filtrable elements on the one hand, and on the other, the "normal" Shiga culture (S type). The G type culture therefore takes in a certain range of culture development, and the forms embraced within it may, accordingly, be somewhat variable, depending on whether they lie nearer to the filtrable elements or to the "normal" culture. This variability is manifested not only in cultural features, but also in the morphology and staining reactions of the individual elements.

direct cultivation to other agar plates, as stated, often did not succeed. When colonies were transferred to broth, it sometimes happened that the resulting growth was abundant after forty-eight hours and characterized by homogeneous clouding. In other cases, the growth was agglutinative, in others faint and delicate and in still others either absent or invisible. The growth in sealed ampules of G filtrates, after standing for from several months to two years, was composed of a viscogranular sediment accompanied by only slight clouding of the medium. Regardless of these variable appearances of the G cultures, it was usually not difficult to perpetuate the forms by broth to broth transfer.

Probably one of the most striking characteristics of the growth in broth seeded with G colonies that had first appeared on agar plates was the irregular results relating to the time when growth first appears and to the ultimate degree of growth as indicated by the amount of clouding or of sediment. One might subculture from half a dozen colonies all of which looked alike and had approximately the same size; but there might be no two tubes that would manifest exactly the same characteristics at the end of twenty-four or forty-eight hours. In some there might be no growth whatever; in others it might be exceedingly faint and delicate in its clouding; in others the growth might be of a finely granular or sedimentary type, and in still others it might be more luxuriant than in the ordinary culture of the S type Shiga. These cultures in broth sometimes also behaved differently on subculture. For instance, one that showed only a faint clouding might yield numerous G colonies on the plate. Another that showed a heavy turbidity might fail to yield a single colony, even though several drops of the culture were placed on the agar plate. Another might yield, not only the minute G colonies of various grades, but also some colonies that were intermediate between the G and the S form. These observations serve to convince one that, even though the G colonies manifest in themselves a marked resemblance on the plate, their potentialities for development in broth are very different; they indeed comprise a highly heterogeneous group in the ontogeny of the species.

When G broth cultures were passed through Berkefeld candles of N or W grade, the results were different with respect to the form of growth that arose, both in the filtrates themselves and in broth tubes inoculated either with the filtrates or with the "growth sediments" deposited in them. In these cases visible growth was invariably long delayed, often being absent for weeks or months. When broth tubes were inoculated with small amounts of filtrate, the result was usually negative. When growth did appear, it was usually limited to a faint opalescence or to a faint viscogranular sediment that first became visible only after many days or even weeks of incubation or storage. The sediments in these cases resembled those seen in old bacteriophage filtrates, from which (as two of us [P. H. and J. K.] have demonstrated in unpublished experiments) living cultures of G forms can frequently be obtained, either directly or after another filtration, provided the proper technic is employed. In these cases, subcultivation to fresh broth succeeded best when a drop or more of the inoculum was drawn by capillary pipet from the faint sediment at the bottom of the ampule after long standing. The inoculation of such sediments into fresh broth gave rise, after a variable time, to similar sediments or to faint opalescence. Homogeneous clouding of the tubes seldom occurred.

Cultures of the G forms in unsealed tubes of plain infusion broth stored at room temperature or in the cold room retained their characteristics of form, colony type and filtrability over many months. Tubes examined after from six to eight months showed no change, while those examined at the end of eighteen months (in the cold room) showed reversion to the S form. G type cultures sealed in glass ampules and kept at room temperature retained their characteristics for more than two years, and from such cultures the "normal" form of the Shiga culture can be obtained by further passage [10] (see section on viability of G forms, p. 87).

MORPHOLOGIC CHARACTERISTICS OF G FORMS

The appearance of the G forms on their first growth in broth was much the same in all strains examined, but with continued cultivation, either in broth or on agar, changes usually occurred. The most common morphologic unit was a minute gram-negative coccus the size of which varied from the lowest limit of microscopic vision (with 1/12 oil immer-

---

10. As will perhaps be clear from the following sections, it seems reasonable to conclude that the Shiga species is not more "short lived" under conditions of artificial growth than most other intestinal organisms. What makes it appear so is the circumstance that this species, somewhat like the cholera vibrio, possesses unusual facility in passing over into a latent stage of development that has not been commonly recognized by bacteriologists and is not readily cultivable by common laboratory methods. In reality, the Shiga cultures are difficult to kill by lack of care.

sion lens and 10 ocular) to about 1 micron. Forms 0.2 micron in diameter were common. At the beginning, and especially in those cultures showing agglutinative growth, the cocci were arranged in streptoformation or in small clusters. The chains measured on the average from 4 to 11 microns in length, but sometimes exceeded the diameter of the immersion field. In broth, the changes in the morphology of the G units were usually accompanied by corresponding changes in the macroscopic appearance of the growth; the agglutinative sediments usually revealed the streptoforms or the cluster forms, while the faint homogeneous cloudings showed free cocci and granules.

The progressive changes in morphology accompanying further propagation involved at first an elongation of the coccus forms, with some increase in size. Such cultures streaked on agar gave typical G colonies containing round or coccoid elements. These cells often contained one or two red-staining granules (Giemsa stain) at the poles or diametrically opposite to one another in the coccus forms. These cells appeared to be identical with the so-called "dimychit" and "mychit" pictured by Enderlein and described, but not named, by a few other investigators. From these bodies in the course of time the rod forms typical of the Shiga culture developed.

Microscopic examination of stained films from the minute colonies first appearing on agar plates was carried out by three methods: (1) direct transfer of the growth to slide or cover glass, (2) impression smears made on cover glasses by the usual method and (3) impression smears made by the method of Kühn and employed by him in his study of the Pettenkofer bodies. This, in brief, involved placing a cover glass on the field (agar surface) to be examined, cutting out the block of agar with a cork borer of appropriate size, fixing the film without removing the cover glass by a potassium bichromate mixture acting through the agar and finally removing and staining the cover glass by the Giemsa method.

Examination of the G type colonies by these methods revealed elements that resembled in considerable degree those seen in the broth cultures. Both streptococcus and staphylococcus groupings appeared. The size of the granules varied from the limit of vision up to 1.5 microns. The majority had a diameter of from 0.3 to 0.5 micron when seen by the aid of a 1/12 immersion lens and a no. 10 ocular. Besides these bodies, however, there were peculiar sheathlike structures, sometimes containing minute cocci or granules and at other times apparently empty of contents. The sheath material stained light pink by the Giemsa stain, while the granules were purple. There were also peculiar mycelial networks, sometimes made up of well stained threads (bluish or purplish), at other times of poorly stained filaments. In the length of these, minute cocci or granules could often be observed. Finally, forming a background for these various elements, we could frequently detect a curious filmlike substance of viscous or gelatinous consistency in which the coccus elements seemed to be embedded. It may have been an artefact, but it was not observed in the control preparations. As will be pointed out later, the coccus elements were seen especially in the "films" that developed on the surface of certain agar plates inoculated with the G culture filtrates before the actual G colonies made their appearance.

SUMMARY OF SECTION AND CONCLUSIONS

From the data presented in this section it will be apparent that we have been dealing with a form of culture possessing somewhat unusual characteristics both in morphology and in cultivation. The presence of the peculiar granular bodies, coccal elements, streptococcus forms and peculiar sheathlike structures might make it appear unrelated to the Shiga culture of origin. The idiosyncrasies of its behavior in continued propagation in broth and on agar plates, its appearances and disappearances from the mediums, suggest a form of growth that, to say the least, is not familiar to bacteriologists. And yet it is of interest to note that some of these characteristics appertain to the G type cultures obtained from other related bacterial species. Moreover, the G type cultures are not unlike many of the peculiar forms that have been described for some of the earliest of the "antebacillary" stages of the tubercle bacillus (Panek and Zakharoff). Some of them possess a striking resemblance to some of the bodies that have been pictured for the virus of contagious pleuropneumonia of cattle. It might suggest to us that these elementary stages of development of bacteria meet on a common morphologic basis, and that it is only or chiefly in the more highly organized, mature forms that the distinctive and characteristic features of the species are manifested.

On the other hand, the results obtained thus far might indicate nothing more than that we are dealing with a contaminating bacterial form which possesses some unusual characteristics, which has, in some unknown manner, gained access to our Shiga cultures, and which has been perpetuated with them. This, then, would be a contamination sometimes present in such small numbers that it evades discovery on the plates and at other times so numerous that it appears to the exclusion of all other culture elements. At present, we shall not attempt to answer this possible objection, but pass on to consider other interesting characteristics of this new culture type.

BIOCHEMICAL REACTIONS, SEROLOGIC REACTIONS, VIRULENCE AND TOXICITY OF G TYPE CULTURES

The subjects here concerned have been studied only in a preliminary way. But, in order to afford some degree of completeness in this work, it seems desirable to incorporate such data as are available at the time of writing.

BIOCHEMICAL REACTIONS

The G strains employed in the fermentation tests had all passed through the virus stage on one or more occasions and had eventually been brought back to the G form, in which they appeared well stabilized at the time of study. In addition, a subculture of one of the strains had been led back to the "normal" Shiga type, thus confirming the integrity of the cultures used. At the time of the tests, the colony form was the same as that of most of our G strains. The diameter was from 0.05 to 0.15 mm. The organisms themselves were predominately cocci or granular bodies with a tendency to streptococcus arrangement. This feature showed in the colonies and was intensified in the sugar mediums, as will be pointed out later.

Only fermentation tests were employed. The sugars selected were those commonly fermented by the three chief dysentery types, Shiga, Flexner and Hiss; they were dextrose, maltose and mannite, purified by filtration through Berkefeld N candles and added to sterile, fermented infusion broth medium of p_H 7.6. The G strains used were 28Aa, 34Ab, 35Ac and 21Ac. The growth was fair in all of the mediums, except mannite. In two of the strains the growth, at the end of twenty-four hours, was heavier than that ordinarily observed in "normal" Shiga (S type) cultures (34Ab and 35Ac). The readings were made at the end of twenty-four and forty-eight hours by means of the potentiometer. At the end of the tests, plates were streaked with each culture in each medium; in addition, stained preparations were made for the purpose of ascertaining whether the sojourn in the sugar mediums had caused any modification in the morphology of the cells. In addition to the aforementioned tests, tubes of litmus milk were inoculated and examined at the same time as the sugar mediums. The results of these tests are presented in table 8.

From table 8 it appears that, of the four G strains studied, two fermented both dextrose and maltose, while none fermented mannite; in fact, two fermented none of the sugars employed. It may be noted that the original Shiga culture fermented dextrose, but was negative toward maltose and mannite. It is clear that the G strains failed to show the essential fermentation characteristics of the Shiga, the Flexner or the Hiss-Russell dysentery types. The failure of the first and fourth G strains to ferment dextrose more definitely, although growing well in this medium, is peculiar.

TABLE 8.—*The Fermentation Reactions of Four Strains of the Shiga G Type Cultures on Dextrose, Maltose and Mannite; Also Reaction in Litmus Milk; Also the Morphology of the Cells and Their Cultural Appearance at the Termination of the Tests*

	Dextrose*			Maltose*			Mannite*			Litmus	Agar	Stained
Strain	24 Hr.	48 Hr.	Gr†	24 Hr.	48 Hr.	Gr	24 Hr.	48 Hr.	Gr	Milk	Plate	Preparation
28Aa	7.4	6.8	3+	7.5	6.8	3+	7.5	7.3	2+	No change	G colonies	Fine streptococcus forms and granules
34Ab	5.2	...	4+	5.8	...	4+	7.4	7.4	2+	No change	G colonies	Fine streptococcus forms and granules
35Ac	5.0	...	4+	4.9	...	4+	7.4	7.4	2+	Acid	G colonies	Fine streptococcus forms and granules
21Ac	7.5	7.2	2+	7.5	7.4	2+	7.6	7.5	2+	No change	G colonies	Fine streptococcus forms and granules
S type	5.3	5.2	4+	7.3	7.1	2+	7.4	7.3	2+	No change	S colonies	Short rods

* Initial reaction p_H 7.5.
† Gr = degree of turbidity of broth.

The failure of these G strains to reveal the type fermentation characters of the Shiga culture may be regarded by some as militating against their inclusion within the Shiga species. We do not regard the matter in this light. It is becoming increasingly evident that different cyclostages within the ontogeny of a species possess varying fermentation characteristics, and some are likely to be different from the textbook cultures. Indeed, d'Herelle (1930) well showed how different may be the biochemical features of different cultures arising from different colonies of the bacteriophage-resistant form of the cholera vibrio. Moreover, Miss Richardson and one of us (P. H.) found similar discrepancies in the fermentation reactions of different G cultures of the diphtheria bacillus in their early stages of development.

Litmus milk was changed in only one instance. Here there was a slight production of acid.

The morphology of the organisms arising from the sugar mediums after the completion of the tests was of interest. The predominant form was a fine-chained streptococcus, similar to the form seen earlier in the same cultures. There was slight variation in the different G strains in this respect. All of the strains, when plated after the completion of the tests, revealed minute G type colonies similar to those present at the beginning. None of these G cultures have yet been transformed to the mature Shiga type, but we anticipate that it will be possible to accomplish this, as it has been done in other cases and, in fact, with a subculture of the same G strain as it stood in 1928.

Of final interest in the biochemical reactions of the G strains is the behavior of the strains that "reverted" to the original culture form. G strain I, as previously indicated, reverted to a small R type, while strain II reverted to an S, after from fifty to sixty passages in infusion broth. At this time they fermented dextrose, but not maltose or mannite. They reduced nitrates slightly, but did not produce indol, ammonia or hydrogen sulphide. In addition, they were again susceptible to the Shiga bacteriophage.

RELATION TO OXYGEN

G strain I was inoculated in a plain agar plate and grown in the presence of 0.2 per cent oxygen. The colonies developed to the size shown by the control, but they were somewhat more flat. Four other G strains were grown in deep tubes of liver broth covered with a one-half inch (1.27 cm.) petrolatum seal. The usual degree of clouding of the medium occurred.

THERMAL DEATH POINT

Preliminary observations on a culture of the cholera vibrio had suggested that the G forms had a higher thermal death point than the S type culture. For this reason, we tested the G forms of Shiga in comparison with the S type cells. Cultures were made in broth of p_H 7.4. The capillary tube method was employed, and the temperatures used ran from 54 C. to 60 C. Four capillary tubes of each sample were tested. The results showed that the G forms were unable to produce colonies on plates after heating at 55 C. for ten minutes. The same was true of the S type culture. Whether the virus form in these cultures was destroyed by the same treatment cannot be stated, since there was no opportunity to carry out serial agar plates on the heated tube samples. The tubes were, however, kept under observation for two months and at the end of this time showed no sign of culture growth. They are still under observation.

SEROLOGIC REACTIONS

Antigen was prepared by centrifugating in Hopkins tubes large amounts of broth culture of the G forms (strain I). The strength of the suspension was the equivalent of 0.01 cc. of packed sediment to 6 cc. of salt solution. The antigen was not killed. Five intravenous injections of 1 cc. each were administered. The rabbit so inoculated showed no unfavorable symptoms other than a loss of weight amounting to about 300 Gm., from which recovery was made. Eight days after the last injection, the rabbit was bled by cardiac puncture and the serum tested against the homologous antigen and against S and R type antigens, by agglutination. For comparison, an anti-S Shiga immune

TABLE 9.—*The Agglutination Reactions Between Anti-G, Anti-S and Anti-R Immune Serums and G, S and R Antigens*

Antigen[a]		Serum	Anti-G Serum Dilutions								
			20	40	80	160	320	640	1,280	2,560	Control
G	....	G	C	C	C	C	3	2	1	—	—
S	....		—	—	—	—	—	—	—	—	—
R	....		—	—	—	—	—	—	—	—	—
S	....	S	C	C	C	C	C	C	4	2	—
R	....		C	C	C	C	C	C	C	3	—
G	....		1	—	—	—	—	—	—	—	—
R	....	R	2	1	—	—	—	—	—	—	—
G	....		2	1	t	—	—	—	—	—	—
S	....		2	1	—	—	—	—	—	—	—

[a] All antigens were in living state.

serum and an old anti-R serum were also employed in cross-tests. Antigen and serum dilution were used in amounts of 0.5 cc. each. The results are presented in table 9.[11]

From table 9 it appears that no relationship was shown between the G and the other antigens. The S serum agglutinated well both the S and the R antigens, as was to be expected. The R serum, on the other hand, had little effect on the other antigens or even on the homologous antigen. This serum was 3 years old at the time of use, and this may be one explanation of the poor results. The most important fact is that the G serum had no effect on the S or on the R antigens, while agglutinating well its own antigen.

---

11. The centrifugated G antigen formed such a compact mass in the Hopkins tubes that it was necessary to remove it and shake for a considerable time with glass beads in order to produce a homogeneous suspension.

Of final interest in the serologic reactions involving the G type cultures is the behavior of the G antigen after "reversion" of the culture to the original type (S or R). Among the G strains that so reverted was strain I, which, after many passages, gave place to a culture characterized by a small, but otherwise typical, rough colony; also G strain II, which, after many passages, reverted to the smooth type culture. Antigens were prepared for these two strains and also for a G strain that had not reverted, but remained well stabilized. These three antigens were tested against an anti-S Shiga immune serum. The results were as shown in table 10.

From table 10 it appears that the anti-S serum agglutinated the R (reverted) antigen well, the S (reverted) antigen poorly and the G (stable) antigen not at all. It might be regarded as strange that the S serum should agglutinate an R antigen better than its own; but, in

TABLE 10.—*The Agglutination Reactions of Two Reverted G Strains and One Stable G Strain in the Presence of an Anti-S Shiga Immune Serum*

Antigen	Anti-S Immune Serum Dilutions								
	20	40	80	160	320	640	1,280	2,560	Control
R (reverted)	C	C	C	C	C	C	3	2	—
S (reverted)	—*	3*	4*	C	C	2	1	—	—
G (stable)	—	—	—	—	—	—	—	—	—

* Apparently a prezonal action.

reality, this is what we should expect. This apparent discrepancy in agglutination occurs frequently and is perhaps the rule among these dissociates. On the other hand, the pure G antigen gave no sign of reaction in the S-immune serum. The S antigen manifestly revealed a prezone action.

TOXICITY AND VIRULENCE

Only a brief study of this aspect of the problem has yet been made. Two rabbits were inoculated with successive doses of the living G type cultures for the purpose of securing an immune serum.

One received intravenously five doses of 1 cc. each of a living concentrated (centrifugated) G antigen from G strain I. The doses were given on alternate days. Aside from a loss of weight of about 300 Gm., no unfavorable signs developed, and the animal remained well until bled eight days after the last dose.

Another rabbit received intravenously three doses of from 1 to 2 cc. each of a living concentrated G antigen from G strain III, for the purpose of developing protection against the toxic S type Shiga culture. The doses were given on alternate days. Two weeks after the last injection, this rabbit received 0.1 cc.

of living S type Shiga culture intravenously and survived without any apparent unfavorable symptoms. A control rabbit receiving the S type culture alone died in from twenty to twenty-four hours.

These tests demonstrate the harmless nature of the living G type Shiga cultures administered intravenously in considerable amount to rabbits. One experiment suggests some degree of immunizing power in the G type antigen, although this single instance scarcely warrants any conclusions on this point.

SUMMARY OF SECTION AND CONCLUSIONS

These brief observations on the biochemical and serologic characters of the G type cultures obtained after recent filtrations indicate how little they agree with the chief features of the ordinary Shiga culture. In fact, by the tests employed, the G forms could not be recognized as belonging to the Shiga species at all. As will be shown later, however, the G cultures may be caused to revert to the original type and to acquire all of the characters of the species, although the time involved may be long. These results may perhaps be accepted as evidence that the specific characters of bacteria, as usually recognized and catalogued by bacteriologists, are in reality representative of the species only at certain stages of its ontogeny, and that these stages are more likely to represent the phase of maturity than the period of physiologic youth.

EXPERIMENTS ON FILTRATION

In the section dealing with the origin of the G type cultures it was shown how these arose in association with, or following, the disappearance of the "normal" Shiga cultures; later sections have revealed something of the peculiar morphologic and cultural characteristics of this new culture type. Although the possibility of causing these elements in the G cultures to traverse Berkefeld candles was mentioned on a previous page, the actual proof of their constant and easy filtrability, and the details of the methods employed, remain to be presented.

As noted earlier, it was not our intention in these investigations to attempt to produce the filtrable forms by filtration procedures. On the contrary, it was our intention to produce a new form of the Shiga culture by other means; then, if possible, to demonstrate its filtrability. As intimated in the introduction to this paper, we had ample grounds to suspect the existence of such a culture type. On the basis of evidence presented, we cannot conclude that we have actually produced such a culture type. Let us say, for the present, merely that we have found associated with the Shiga cultures on numerous occasions and in certain stages of development a new culture form, which, for convenience in reference, we have termed the G type culture. Perhaps it is a contamination; perhaps it is not. Without discussing this point further at present, we may turn to our records, which demonstrate that, whatever may be the nature of the G forms, they (or elements accompanying them) are readily and invariably filtrable through all grades of Berkefeld candles, and that they can be recovered from the filtrates in nearly all instances by the employment of a suitable cultivation technic. No trace of them can be found, however, by ordinary cultivation methods, except sometimes in the V filtrates and less frequently in the N filtrates.

GENERAL CONSIDERATIONS AND TECHNIC

In the experiments on filtration now to be reported, grades V, N and W of the Berkefeld series were employed. Some of the candles were old, some had been used a few times, but the majority were new. The age of the candle or the amount of usage seemed to make no difference in the results obtained. During the latter part of the work all old candles and many of the new ones were tested by air pressure; no N or W candles were employed unless they could withstand an air pressure of more than 9 pounds (4.1 Kg.) without showing bubbles in the candle proper or around the metal base. Many old candles were thrown out on these tests. We believe that the air test is of value in revealing at least the larger leaks. The usual filtration technic, with few exceptions, was maintained. The candles, after being baked and cleaned, were sterilized, with mantle and side-arm tube attached, in the autoclave at 121 C. for about twenty minutes. The temperature was then slowly raised to 134 C. for about five minutes. In some cases, water was placed inside the mantle before sterilization, but in most cases it was not. We were unable to detect any damage to the candles resulting from the use of the high sterilizing temperatures. The ordinary forms (S and R) of the Shiga bacillus did not pass the candle in any case. They are in fact nonfiltrable under the conditions of our work.

Fractional filtration was employed in many instances. The amounts filtered were usually between 20 and 35 cc. The total time required for the filtration of these amounts was between two and seven minutes when negative pressure was used. It was longer with the W and N candles, shorter with the V and much longer with gravity filtrations. Most of the filtration was accomplished under negative pressure. This never exceeded 300 mm.; in the majority of cases, it was only from 150 to 200 mm. A small number of filtrations were made on gravity. The results were positive in nearly all the tests, whether with gravity or with negative pressure. The filtrates were collected in the sterile side-arm tubes attached to the filter candles and were removed therefrom by means of sterile Pasteur pipets, blown from sterile stock tubing just before use. The filtrates were then distributed into tubes previously sterilized at 210 C. Certain tubes of the fresh filtrate were sealed in glass ampules. These were laid away in a closet at room temperature for the purpose of revealing possible growth or contaminations. As will be described in the section on viability of the G forms, these sealed ampules were observed and examined culturally more than two years later. For the fractional filtrations, in some cases, a modification of Mudd's filtration assembly was employed. Samples of filtrate were secured at intervals of from thirty seconds to one minute and then placed in separate tubes to await further study.

For filtration purposes, the G type cultures, prepared from single colonies by "fishing" with glass threads, were grown in beef infusion broth of p_H 7.6 for from twenty-four to forty-eight hours at 37 C. When such G cultures, which ordinarily showed only a faint clouding, were filtered through Berkefeld candles of any grade, some of the elements in the culture almost invariably passed the candles. The presence of these filtrable bodies was demonstrated in three ways: (1) by growth in the original filtrate after incubation in unsealed tubes; (2) by growth in sealed ampules after a period of from several months to two years or more, and (3) by serial plate cultivation of the fresh filtrates on plain infusion agar, litmus lactose agar (Hauduroy's method, 1927, a) or blood agar—the last being seldom used.

When growth first appeared in the samples of tubed filtrate, it was invariably long delayed—sometimes for weeks or months—and seldom gave a distinct clouding of the medium. At the end of several months there was often an opalescence and usually a viscogranular sediment. From such tubes, whether showing opalescence or sediments, or remaining clear, the G type culture could eventually be recovered by the serial plating method soon to be described. In the case of sealed filtrates two or more years old, the G colonies often arose on the first plate after an incubation at 37 C. for from six to ten days, as will be shown in detail in the section on viability of the G forms.

The seeding of from 5 to 10 drops of fresh filtrate on a sterile agar plate usually resulted, after two or three days' incubation, in what seemed to be a sterile plate; only a delicate "film" could be discerned. Even under microscopic examination nothing that resembled colony structure was revealed. When this film was washed up in from 10 to 15 drops of sterile broth and transferred to a second plate, this plate also remained apparently sterile, although, like the first, it presented a film-like surface. On the third, fourth, fifth or later plate, however, similarly inoculated from the plate just preceding it in the series, it was possible to detect by a hand lens (3.5 × to 10 ×), a no. 3 objective, or even by the unaided eye the very minute and delicate colonies representing the G type culture, apparently taking form from the filmlike débris. Continued plate transfers in series at frequent intervals served to maintain the culture in this form for many generations. This is essentially the method first employed by Hauduroy (1927, a) and, as in the cases described by him, the continued plating rendered the growth more stable on the solid medium. We did not, however, observe the rapid evolution of the G type culture into the original form, that he reported—at least, when the G culture was derived from fresh filtrates. As will be pointed out later, the "reversions" noted by us occurred much more slowly.

When the fresh filtrate was permitted to stand at room temperature for some days or weeks and then was inoculated in agar plates, the G type colonies arose more quickly, sometimes on the first plate seeded. This was especially true of the V filtrates. In these cases, the minute colonies were often numerous, although the filtrate itself might reveal no clouding. It would seem as if some process of development of the filtrable elements occurred in the standing filtrate; forms, at first non-cultivable on agar, slowly became cultivable. This possibility will be referred to again on a later page.

FRACTIONAL FILTRATIONS

The rapidity with which the G colonies appeared on plates seeded with fresh filtrates or in the reserve tubes of the filtrates themselves is indicated further by the following data, derived from some of the experiments on fractional filtration. In addition, these data have a bearing on the question as to the exact time in the filtration process when the filtrable elements pass the candles most readily. In table 11 are presented the results of five series of tests, each involving the use of a culture that was filtered in three or four fractions. The negative pressure was interrupted after the passage of each sample. In case, after two platings from a given filtrate fraction, no sign of G colonies appeared, the reserve filtrate of the same fraction was examined for growth. Whether this examination of the reserve showed growth, doubtful growth or no growth is indicated by appropriate symbols in the column giving the results of the plate tests. The symbol F should be interpreted as signifying the presence of the filmlike substance on the surface of the inoculated agar. The footnotes attaching to the series numbers of the samples filtered present additional and essential data relative to the conditions surrounding the filtration of each series.

From table 11 the following points may be noted. In one series only (5) did the G colonies first appear on the first plate seeded. Moreover, in no other series except one (series 1, fraction 2) did the G colonies appear in the second serial plate. Thus, to summarize the positive picture (omitting series 5), these colonies appeared for the

TABLE 11.—*The Time of Appearance of the G Type Colonies on Plates Inoculated with Fractional Filtrates of G Type Cultures in Experimental Series 1 to 5; Also the Presence or Absence of Visible Growth in the Reserve Tubes in Case the Plates Were Negative*

Plate	Time Out of Filtrate, Hours	Series 1* Fractions			Series 2† Fractions			Series 3‖ Fractions (Gravity)				Series 4‡ Fractions (S Type Culture)				Series 5¶ Fractions (Strain V)			
		1	2	3	1	2	3	1	2	3	4	1	2	3	4	1	2	3	4
1	48	F	F	F	F	F	F	F	F	±	F	F#	F#	F#	F#	+	+	±	+
2	96	F	+	F	F	+	F	F	±	F	F	F#	F#	F#	F#	+	+	+	+
3	144	F	+	F	F°	—°	—°	—°	—°	—°	—§	±§	±°	+°	±§	+	+	+	+
4	192	+°	+°	F°	F	—	—	—	—	±	+°	÷#	F°	+°	+°				
5	240	+	+	F	F	—	—	—	—	+	+	+	—	+	+				
6	296	+	+	—	F	—	—	—	—	+	+	+	F	+	+				
7	336	+	+	...	F	—	—	—	±	+	+	+	F	+	+				
8	384	+	+	—	+	—	—	—	+	+	+	+	+	+	+				

F, film on surface (no colonies); ±, doubtful film; +, distinct colony formation (G type); —, no film and no colony formation.
° Growth in reserve filtrate tube on this date.
No growth in reserve filtrate tube.
§ Growth in reserve tube doubtful.
* Berkefeld N candle, new; negative pressure, 100, 120 and 150 mm.; three fractions at intervals of thirty seconds, one minute and one minute; total filtration time, two minutes and thirty seconds; total volume, 30 cc.
† Berkefeld N candle, new; negative pressure, 180, 220 and 220 mm.; three fractions at intervals of thirty seconds; total filtration time, about two minutes; total volume, about 20 cc.; no growth in any medium in first ninety-six hours.
‖ Berkefeld N candle, new; filtration by gravity; four fractions at ten, thirty and fifty minutes and one hour; total filtration time, about two hours and thirty minutes; total volume, 30 cc. Results: fraction 1, negative plates and positive reserve at twenty days; fraction 2, positive plate (eighth) at sixteen days; fraction 3, positive plate (fifth) at ten days; fraction 4, positive plate (fourth) at eight days.
‡ Berkefeld N candle, new; filtration of "normal" S type Shiga broth culture; negative pressure, 140, 180, 180 and 180 mm.; four fractions at intervals of thirty seconds, one minute, two minutes and three minutes; total volume filtered, 30 cc. Results: fraction 1, positive plate (fourth) at eight days; fraction 2, positive plate (eighth) at sixteen days; fraction 3, positive plate (third) at six days; fraction 4, positive plate (fourth) at eight days.
¶ Berkefeld N candle, new; G type culture, strain V; negative pressure, 120, 120, 120 and 140 mm.; four fractions at intervals of thirty seconds, one minute, one minute and two minutes. Results: first plate of all fractions positive; fractions 1, 2 and 4 positive after forty-eight hours; fraction 3, after seventy-two hours. This is the only series in the present case in which the G colonies appeared on the first plates seeded. They were of the typical G type.

first time: once on the second plate in the series, once on the third plate, four times on the fourth plate, once on the fifth plate and twice on the eighth plate. Of the fourteen fractions included in these experiments (omitting series 5), four failed to yield visible colonies by the eighth plate, although all of these had shown, at some stage of the plate transfers, the peculiar surface film mentioned earlier. In all of these instances, however, the presence of living organisms was demonstrated in the reserve samples of the filtrates after a sufficient time had elapsed. In one instance only (series 4, fraction 1) did a filtrate that yielded either the surface film or the colonies, or both, fail to manifest recognizable growth in the reserve filtrate tube.

The time element involved in filtration tests has often been of special concern to bacteriologists. In the present experiments, the time limits employed seemed to play no appreciable part in the results. The filtrable elements associated with the G type cultures passed the candles as readily during the first minute of filtration as during the last, or as during any intermediate minute; and this was true irrespective of the volume filtered, the total filtration time or the amount of negative pressure employed. We have no doubt that the time element is of importance when an investigator is attempting to force through a candle the common, much larger culture elements, such as the cells of the S and R cyclostages. These, we believe from our experience, are not filtrable in the sense in which this term is employed in the present work. However, to the time of writing, we have not been able to discover a means for preventing the passage through the filter candles V, N or W, of the minute elements present in all cases in the G type cultures.

One other point brought out in table 11 is of special interest and deals with fractions 1, 2, 3 and 4 of series 4, involving the filtration of a broth culture of the "normal" or S type Shiga stock, which, to the time of this experiment, had not been regarded as harboring the filtrable elements. It is indicated by the tabulation that none of the four fractions of the filtrate, tested by the plating method, revealed definite G colonies in either of the first two serial plates. Three of the four, moreover, failed to reveal colonies on the third plating. At this moment one fraction only (3) showed colonies. Two of the four fractions gave the first evidence of living forms only in the fourth serial plate. One (fraction 2) revealed no clear evidence of colony development until the eighth serial transfer. It is thus apparent that, although the original S type culture gave no gross appearance of being other than the "normal" type, the G elements, or the filtrable forms that precede or accompany them, were present within it. In no case, however, was an S type colony observed on the plates. In this instance it therefore appears that filtration served to separate the G form in purity from the culture of mixed (G and S) type. In this instance, however, we unfortunately have no evidence that the G elements were actually present with the S before filtration was carried out. This case, moreover, is the only one in all of our work in which it might be said that we "produced" the G form by resort to filtration; and, as we have noted, this instance was inadvertent.

In the filtration tests outlined it will be noted further that in each case a new Berkefeld candle was employed. Apparently the same order of results was obtained by the use of candles that had been in service for some time. In work of this character we therefore see no reason to insist on the use of new candles, at least for routine procedures.

One important aspect bearing on any conclusions as to the existence of distinctly filtrable stages in the life history of bacterial species, as opposed to stages that are not filtrable, is the number of elements that pass through the candles and are present in the filtrates. As pointed out by Mudd, the observation of a colony or two on a plate seeded with a considerable amount of filtrate is not, in itself, a striking or significant phenomenon. If, on the other hand, a filtrate presenting the optical appearance of sterility reveals hundreds or thousands of filtrable and cultivable bodies for every drop, one has better reason to suspect in the species concerned the existence of elements that could appropriately be described as "filtrable forms."

In consideration of this question, it will be apparent that, in our tests, we could not, as a rule, safely employ either the N or the W filtrates, for the reason that, in this event, there would have existed no basis for estimating the number of living elements that renewed their visible growth on the first plate streaked with the filtrate. This was because, as explained, the first plate of the series usually yielded no G colonies; and the number of colonies that appeared on the second, third, fourth or later serial plate did not indicate the number of living elements that were present on the first. For this reason, it was chiefly the V filtrates that could be utilized for this study. For, with use of these, as noted, the G colonies more frequently appeared on the first plate inoculated with the filtrate. In other words, serial transfers from plate to plate were not required to bring the filtrable bodies into colony development. In a few cases, however, the N filtrates also yielded distinct G colonies on the first plate, and these instances also gave some information of value. But it was mainly from a study of the V filtrates that we obtained the clearest picture of the frequency of occurrence of these filtrable bodies associated with the G type cultures which were able, without appreciable delay, to reproduce in colony form. Naturally, such a method did not reveal the number of filtrable elements that passed the V candles, but were unable to produce at once the colony picture of the G type culture.

Most of our early work with the V filtrates involved the use of large amounts—usually from 5 to 15 drops—of filtrate, seeded to the agar plates. The results were variable and seemed to depend on the stage of development of the G culture. In some cases, when the G colonies first appeared, there were no more than a dozen that were visible with a hand lens or to the unaided eye at the end of forty-eight hours' growth, although more could usually be detected with the no. 3 objective. In other cases, seeding the plates with 5 drops of the V filtrate gave hundreds or even thousands of colonies, the majority of which were probably visible with the unaided eye. Occasionally similar results were obtained with an N, but never with a W, filtrate.

In one case, a V filtrate was collected in seven fractions, taken at intervals of thirty seconds, and 0.5 cc. of each fraction was seeded on an agar plate. After twenty-four hours' incubation, plate 1 gave 20 visible colonies (unaided eye), plate 2 gave 200, plate 3 gave none, plate 4 gave 10 and plate 5 gave none. After thirty-six hours' incubation, however, plates 3 and 5, previously negative, each showed more than 100 colonies, and the number was increased proportionately on the other plates. There were occasional instances in which the seeding of one loop of a fresh V or N filtrate yielded, after from thirty-six to forty-eight hours, many hundreds of G colonies.

These results with the V filtrates, and occasionally with the N, indicate that the filtration of these elements does not necessarily involve the passage through the candle of a few elements only, but of hundreds or thousands for each cubic centimeter of filtrate. The bodies being considered are such that they are able to express themselves at once as G colonies. How many bodies that cannot manifest directly such visible colony development, but which may yield G colonies after repeated passage on agar plates, has not been ascertained; and we do not see that it can be—by reasons of the conditions involved. If colonies do not appear on the first plate seeded, the number appearing on a later plate possesses no significance for estimating the total number of filtrable bodies that passed the candle. For the sake of completeness, however, this matter may receive brief further consideration.

If no G colonies appeared on the first serial plate, this was washed up with about 10 drops of sterile broth spread over the surface by means of a sterile, bent, glass rod. Of the 10 drops added, about 3 or 4 drops could usually be recovered, and this amount was spread over the surface of a second plate. If no colonies appeared on this plate, the process was repeated, and so on through a longer or shorter series. When, on some plate of the series, the G colonies finally appeared, the number was great, ordinarily from 2,000 to 100,000. In some cases the plate surface showed a solid film of growth. Successful inoculation of later plates could usually be carried out by touching the platinum needle to one or more colonies and then inoculating several drops of sterile broth, previously placed on a fresh plate, then spreading. As reported on another page, however, direct plate to plate transfer by the "dry" method usually did not succeed until the G cultures had become well stabilized.

From the data presented on the last few pages, it is apparent that when we have referred to a "filtrable form" associated with the Shiga bacillus we have referred to a type of culture that was almost invariably filtrable through all grades of the Berkefeld candle, and that passed, not to the extent of a few individuals, but in great numbers, often crowding the test plates with multitudes of minute colonies, some of which had to be measured, not in millimeters, but in microns. Again it may be stated, however, that these filtrable forms were not "produced" by filtration, for they were first present and demonstrated by plating methods in the original culture stuff submitted to the filtration process.

RAPIDITY OF DEVELOPMENT OF G COLONIES FROM FRESH FILTRATES

The rapidity with which the G type culture developed on agar plates, in the form of recognized colonies, as opposed to the surface film (which we are inclined to believe is made up of some kind of living elements), was often related to the grade of candle employed for the test, as suggested briefly on an earlier page. In general it was observed that when the V candles were employed, the colonies appeared earlier in the series (most commonly on the first plate), although there were exceptions; and that when the N or W candles were used, the colonies were more likely to appear later, most commonly on the third, fourth or fifth plate. Occasional records were kept of the time of appearance of the G colonies; the results in a few of the experiments are assembled in table 12.

From table 12 it appears that in nine tests with V filtrates the G colonies appeared on the first plate in seven instances. Of twenty-seven N filtrates, five gave G colonies on the first plate and three on the eighth plate, but they gave G colonies first most commonly on the second, third, fourth and fifth plates, the largest number giving the colonies first on the fourth. Of five W filtrates, all registered the first G colonies on the third, fourth or fifth plate. The same order of results has been obtained in many other tests with the Shiga G type filtrates, and also by others in our laboratory working with other bacterial species.

RATE OF DEVELOPMENT OF G TYPE CULTURES IN FILTRATE RESERVES

It is a well known fact that when a culture of the Shiga bacillus undergoes so-called "lysis" under the influence of the bacteriophage, and is then filtered through a Berkefeld or Chamberland candle, and the filtrate is set aside for some weeks or months, an opalescence or sometimes a faint clouding often appears. In other cases there occurs a slight sediment. These phenomena are due to the generation of a

TABLE 12.—*The Relation Observed Between the Grade of Candle Employed for Filtration and the Time at Which Recognizable G Colonies First Appeared on the Serial Plates Seeded with the Filtrates*

| Grade of Candle | Number of Cases in Which the G Colonies First Appeared on the Serial Plate Indicated by Number Below | | | | | | | | |
	1	2	3	4	5	6	7	8	Total
V	7	0	1	1	0	0	0	0	9
N	5	4	4	9	2	0	0	3	27
W	0	0	2	1	2	0	0	0	5

new form of the original culture. The new organisms are usually small coccus or coccoid forms, more resistant, or even extremely resistant, to the bacteriophage, as first pointed out by d'Herelle. In the course of time, however, they may become retransformed into the original culture type. Such bodies manifestly arise from filtrable elements present in the "lysed" culture of origin, and they are, in turn, able to give rise to other filtrable forms before they return to the "normal" type (d'Herelle and Hauduroy).

In view of these observations relating to the development of a secondary culture growth, of a sort, from bacteriophage filtrates, which is now clearly established, it is of interest to observe the behavior of the G type culture filtrates with the lapse of time.

In several filtration tests involving N and W candles, less frequently V candles, it was observed that certain filtrates which failed to yield G colonies on the first plate of the series when the filtrate was fresh gave a fair growth of G colonies after the filtrate had stood at 37 C., or even at room temperature, for a longer period. In other words, the elements present in the fresh filtrates seemed to be noncultivable, while elements present in filtrates six or more days old were more easily cultivable on agar plates. A study was therefore made of the time at which cultivable forms could first be demonstrated in the "ripening" filtrates by the agar plate method. Incidentally, the experiment here reported revealed the origin of one of the G strains from an old, and apparently dead, agar slant culture of the R type Shiga culture.

On Dec. 2, 1927, an agar slant culture was made of an R type Shiga culture, produced earlier from S stock under the influence of lithium chloride broth. This R culture, after incubation, was sealed and kept at room temperature. In May, 1928, a subculture on agar was attempted. This attempt, like subsequent attempts, yielded no visible growth. A heavy inoculation to broth gave, after several days, a viscous sediment, but no indication of growth in the supernatant broth. A still larger volume of broth was then inoculated, incubated and finally filtered through an N candle. Directly after filtration, 5 drops of filtrate were spread over an agar plate, and the plate was incubated for forty-eight hours. No growth resulted. On the next day, a seeding was attempted from this plate to a second. At this time the original filtrate showed a slight viscous sediment, although the supernatant liquid was clear. The second plate also remained apparently sterile. On the third and fourth days similar seedings were made, but both of the plates remained apparently sterile. On the fifth day after filtration another plate was streaked. After thirty hours' incubation, superficial examination revealed a veil-like growth covering the greater part of the area seeded. Closer inspection revealed the presence of minute, round, smooth colonies, which in many areas were confluent. On the free edges of the seeded area, the colonies measured from 0.1 to 0.2 mm. in diameter, were irregular in shape and distinctly rough, simulating minute R colonies. The two forms of colony suggested the descriptions and photographs of Haag's smooth and rough gonidial cultures of *B. anthracis.*

These observations seemed to indicate that the living elements present in the stored filtrates become more cultivable with the lapse of time. That this is true is shown further by the results of cultivation of filtrates that had been sealed in ampules for two years or more (see section on viability of the G forms).

Other tests showed that the appearance of the G colonies on plates seeded with the filtrates of G cultures was dependent not only on the number of plates carried in the series, but also on the age of the filtrate when the first plate was inoculated. In other words, the G colonies could be obtained either by carrying forward from plate to plate the washings of each in turn or by waiting for the filtrate to age a sufficient time. The results of some of these tests involving the G forms of

B. typhosus, *B. paratyphosus* B, *B. typhi-murium* and *B. enteritidis* are presented in the following tables. Manifestly the phenomenon is not limited to the G forms of the Shiga bacillus.

In the case of *B. typhi-murium* I, as ascertained by Miss Smith in our laboratory, the time within which the G forms arise may be very short. One instance which manifested unusually regular results is shown in table 13.

Table 13.—*The Relation of the Time of Appearance of the G Colonies of B. Typhi-Murium I to the Age of the Filtrate and to the Number of Plates in the Series*

Number of Plate in Series *	Growth in Series According to Age of Filtrate When Streaked			
	1 Day	2 Days	3 Days	4 Days
1	—	—	+	+
2	—	+	+	
3	+	+		
4	+			

* The plates in series 2 were streaked from series 1; those in series 3 from series 2, and so on.

Table 14.—*The Relation of the Time of Appearance of the G. Colonies of B. Paratyphosus B to the Age of the Berkefeld N Filtrate and to the Number of Plates in the Transmission Series*

Number of Plate in Series	Growth in Series According to Age of Filtrate in Days When Streaked						
	1	3	4	5	6	8	9
1	—	—	..	—	—	—	+
2	—	—	—	—	—	⊥	+
3	—	+	+	+	+	..	+
4	+	C	+	+	+	..	..

.., no plate made, series discontinued; +, G colonies present; —, G colonies absent; C, plate contaminated.

Here it is seen that positive results were obtained from a Berkefeld V filtrate on the third day, but not on the first or second. The first positive result on the agar plate series also occurred on the third day plate. Aging the plates that at first were negative did not result in the development of the G colonies. It thus appeared that whether the "evolution" of the filtrable elements took place in the broth filtrate or on the plates, about three days were required for the process. This result occurred more rapidly than is usually the case with N or W filtrates.

Somewhat similar, but slower and more irregular, results were observed in a filtrate of the G form of *B. paratyphosus* B, as shown in table 14.

Some further irregularities that may be expected in the time of appearance of the G colonies on plates inoculated with fresh filtrates of G cultures are shown in table 15. The Berkefeld N filtrate here concerned was from a G type culture of *B. enteritidis,* studied by Miss Raven in this laboratory.

In table 15 it is shown that the filtrable elements in the aging filtrate were not able to register directly on the first agar plate until the ninth day after filtration. In the plate to plate transfers there was a registra-

TABLE 15.—*The Relation of the Time of Appearance of the G Colonies of B. Enteritidis to the Age of the Filtrate and to the Number of Plates in the Transmission Series*

Number of Plate in Series	Growth in Series According to Age of Filtrate in Days When Streaked*						
	1	2	3	6	7	8	9
1..............	—	—	—	—	—	—	+
2..............	—	—	—	—	—	+	+
3..............	—	—	+	—	—	+	..
4..............	—	—	..	+	±	..	..

* Symbols as in table 14.

TABLE 16.—*The Relation of the Time of Appearance of the G Type Colonies of B. Typhosus to the Age of the Filtrate and to the Number of Plates in the Transmission Series*

Number of Plate in Series	Growth in Series According to Age of Filtrate in Days When Streaked*						
	1	4	5	9	11	12	13
1..............	—	..	—	+	—	—	—
2..............	..	—	—	+	—	—	..
3..............		—	—	+	—	..	
4..............		..	—	..	..		
5..............			+				
6..............			+				

* Symbols as in table 14.

tion of the G colonies for the first time on the third plate of the third series, the next appearance being on the fourth plate (sixth day) of the fourth series.

In certain cases, the filtrable elements that at first were noncultivable yielded the G colony picture at a later time and then seemed to return to the noncultivable state. An instance of this was afforded in the tests performed by Miss Bailey on the filtrable form of *B. typhosus.* An example is given in table 16.

In this case we have another example in which the filtrable bodies present in the fresh filtrate became cultivable for the first time (directly on agar) on the ninth day after filtration. Moreover, the first time that the G forms appeared on the plate series (plate 5, fifth day series) was also the ninth day after filtration. The chief point that this table is intended to bring out, however, is that, after the first appearance of the G forms on the first plate of the ninth day series, they seemed to disappear again, and were not observed on any of the three serial plates of the eleventh day, on the two plates of the twelfth day or on the one plate of the thirteenth day. Attempts to rediscover the culture in the filtrate beyond this point were not made. To the foregoing observations it should be added that the colonies on the first plate of the ninth day did not appear until three days after inoculation and were then of very minute size. In all such tests it is necessary to keep the plates under observation for from three to six days, at least. As a rule, in these serial transfers, plates that are negative at the end of ninety-six hours remain negative. But this cannot be depended on in all cases, for we have on record instances in which the first colony development appeared after from eight to ten days. In any case, the plate surface should be scrutinized with a hand lens having a magnification of at least 10 $\times$, and the no. 3 or no. 7 objectives called in to interpret doubtful colony-like structures.

One of the most interesting cases of the transformation of the filtrable forms into the G type colonies deals with the production and retransformation of the virus form of *B. diphtheriae* (Park 8 strain) as worked out by Miss Ruth Richardson and one of us (P. H.) in this laboratory in 1929. The virus form, in this instance, had been produced earlier by forcing the dissociation of the parent culture under the influence of lithium chloride glycerin broth, used in serial transfers. The Berkefeld N filtrate here considered was, therefore, a filtrate of the diphtheria G type culture, produced as mentioned. The methods of securing the reversion of the virus forms to the G type were essentially the same as those previously indicated, with the exception of the circumstance that the intervals between samplings of the filtrate, and between the serial platings, were considerably longer. The glycerin agar plates were seeded with 10 drops (0.3 cc.) of the filtrate at each sampling. After a period of development at 37 C., the plates were washed over with sterile broth and the "washings" transferred to the succeeding plate. As will be seen in table 17, the virus form of the diphtheria bacillus is characterized by a very slow evolution into the G forms.

In table 17 it is shown that the diphtheria virus forms transformed into the G type cultures in four of the "runs" of G culture filtrate. Four of the "runs" were discontinued at an early date because of the burden of excessive platings. The G colonies appeared in the successive series on the fifty-fifth, forty-ninth, forty-fourth and forty-first days, respectively, out of the filtrate. This seems to demonstrate that increased time of storage in the filtrate tube was of more importance in hastening the time of appearance of the G forms than was time spent on the plate to plate transfers. It was anticipated that the plate "runs" from the samples of filtrate seeded on the fortieth, fifty-fifth and sixty-fifth days would also yield G colonies, but in this we were disappointed. With the seeding of the sixty-fifth day, the filtrate was entirely used and it

TABLE 17.—*The Relation of the Time of Appearance of the G Type Colonies of B. Diphtheriae to the Age of the Filtrate and to the Number of Glycerin Agar Plates in the Transmission Series*

Number of Plate in Series	Interval in Days	Growth in Series According to Age of Filtrate in Days When Streaked									
		1	4	9	14	19	25	32	40	55	65
1	0	—	—	—	—	—	—	—	—	—	—
2	4	—	—	—	—	—	—	—	—	—	—
3	5	—	—	—	—	—	..	+¶	—	..	
4	5	—	—	—	..	—		+	—		
5	5	—	—.	—		—		+	..		
6	6	—	..	—		+#		..			
7	7	—		—		..					
8	8	—		+§							
9	15	+*		+							
10	8	+		..							

.., series discontinued; +, G colonies present; —, G colonies absent.
* Fifty-five days from time of filtration.
§ Forty-nine days from filtration.
# Forty-four days from filtration.
¶ Forty-one days from filtration.

was necessary to discontinue the series. These results demonstrate how tardy is the regeneration of the virus forms of the diphtheria bacillus, compared with those of the Shiga cultures and other members of the intestinal group. The work on the diphtheria cultures will be reported in detail at a later date.

These observations and others that might be reported reveal the fact that the filtrable elements present in the freshly derived filtrates of G type cultures of various bacterial species seldom exist in a form that is visibly cultivable on agar plates under ordinary conditions of growth until many days have passed. The manifestation of this "latent" period varies with the grade of candle and probably with the state of the G type culture at the time of filtration. It has been our experience that this period is, as a rule, from six to twelve days; but it may be much longer, as we have shown in the case of the filtrate of the diphtheria G type culture. With the passage of this "latent" period, the filtrable elements seem to have entered a stage of development in which they are able, not merely to survive, but to register directly on the plates by the formation of the G colonies. Before this time has arrived, several serial cultivations on agar are necessary to reveal their presence. It should also be recognized that these positive results in plate cultivations succeed in the cases of filtrates that, at the time of first attempted cultivation, may reveal no sign of growth, either by sediment, opalescence or clouding. Moreover, although a fair grade of opalescence or even of clouding is apparent, the immediate cultivation on plates is not necessarily rendered more successful on that account.

It might be considered possible that the reason why fresh N and W filtrates are not usually directly cultivable on plates is that the number of viable units passing the filter is too small; that they may not be contained in the amount of filtrate seeded. That this is probably not true is shown by the fact that one cannot detect on the first plate inoculated, either by hand lens or by the no. 3 objective, any sign of a colony—only the peculiar filmlike substance previously mentioned. Moreover, by the time that the G colonies appear clearly, the film has usually ceased to manifest itself. This might suggest that the film is, in some way, the precursor of colony structure; that it represents the first and only possible form of culture development of which the filtrable elements are capable during the earlier stage of their existence. This view receives some support from our microscopic examination of stained preparations (largely impression films), and especially when these are taken into consideration with the control preparations to be mentioned in detail in a later section. The view is also in harmony with observations made by Hauduroy (1929) and will receive further consideration on a later page. As shown in table 11, however, instances were observed, in which the presence of a film was not followed by the appearance of G colonies.

It might be considered possible that serial cultivation of the filtrable elements on agar plates, in order to raise them to the estate of the visible G type culture, is really unnecessary; that, on the contrary, if the first plate on which the fresh filtrate is seeded were kept for a sufficient time, the G colonies would eventually appear. While the "ripening" of the filtrates themselves over a period of days or weeks or months encourages the development of the filtrable elements, it has not been commonly observed that continued aging on the original plate, containing ample medium and protected against drying, accomplishes this end. We have some evidence that this circumstance involves in part the question of ample moisture on the plate surface. On the other hand, sometimes a plate examined after a period of from five to ten days may reveal colonies that were not detected at the end of forty-eight hours or even later. It is also true that when one is dealing with colonies of microscopic dimensions, a few scattered over a plate can easily be missed.

CONSTANCY OF FILTRABILITY OF G TYPE CULTURES

In dealing with the G culture state, the question arises: Is the characteristic of filtrability a transient feature of the culture, just as the G culture may be regarded as a transient form of the "normal" Shiga culture, or is filtrability a constant characteristic, enduring as long as the G type culture endures?

For a study of this point, the G type culture, strain I, was selected, since this had manifested fair stability in continued propagation. This strain was accordingly grown in a series of tubes holding about 30 cc. of plain infusion broth of reaction p_H 7.6. The transfers from tube to tube in series were made at intervals of forty-eight hours. Occasionally the inoculated tubes showed a slight clouding, but more commonly they remained clear over the forty-eight hour period of incubation. Each alternate tube in this series, beginning with the first, was filtered;[12] this gave thirteen filtrations in a series of twenty-five tube generations, covering a period of fifty days. Cultivations were attempted by the serial plate method as soon as the filtrate was obtained. Ten drops were inoculated on the surface of a sterile litmus lactose agar plate, or on one of plain infusion agar, which was then incubated for forty-eight hours, and afterward kept under observation for several days. Each series of plates (one for each day) was continued until positive results were secured.

In the actual experiments a negative filtrate was never obtained in the twenty-two[13] filtrations involved in the experiment, whether the

12. All three grades of Berkefeld candle were employed in this study. In some cases, only one sample from a given tube was filtered. In other cases, two or three fractions were filtered, each through a candle of different grade. Altogether there were twenty-two filtrations in the thirteen samples. Of these, four were through V candles, fourteen through N candles and four through W candles. Every filtration gave positive results: four on the first plate, two on the second, six on the third, seven on the fourth and three on the fifth.

13. Samples of these filtrates and G cultures were sealed in ampules and stored for two years. At the end of this period, the filtration experiment was again resumed (July, 1930). The results of the further tests will be found in the section on viability of the G forms.

filtrations were performed through V, N or W candles. In every instance, distinct G colonies came into evidence on one of the plates, most commonly on the fourth or fifth of the series. These colonies gave rise to cultures which, on further examination, corresponded in all respects with the Shiga G type culture of origin. The filtrability of this G strain thus persisted uninterruptedly over a period of twenty-five serial passages, or a time period of fifty days, after which the experiment was discontinued [13] for the time being.

These experiments performed with G strain I were duplicated in shorter series of filtration tests with other G strains. All gave the same result: they were invariably filtrable and the G strain could always be recovered. But this point must be clearly apprehended: If the demonstration of the passage of the filtrable bodies through the candles had rested on the discovery of the G colonies on the first, or even second, plate, negative results would have been obtained in sixteen of the twenty-two filtrations. If the demonstration had rested on the first plate alone, negative results would have been assigned to eighteen of the twenty-two filtrations. We are thus led to suspect that the conditions here involved often lie at the bottom of the often reported failures on the part of bacteriologists to obtain proof of the filtrability of bacterial cultures.

SUMMARY AND CONCLUSIONS REGARDING THE EXPERIMENTS ON FILTRATION AND ON THE RECOVERY OF THE G FORM OF CULTURE

We are here dealing with the facts of filtration and not with the theories. The facts presented in the two previous sections concern the behavior, in filtration tests, of the G type cultures, the origin, appearance and cultural behavior of which were described earlier. Whatever the actual nature of these cultures may be, we have seen that they are readily and, we may even say, invariably, filtrable. They are filtrable through all grades of Berkefeld candle, new or old. They pass the candles at any degree of negative pressure and equally well, though much more slowly, by gravity. In the course of fractional filtration, they appear in the first fraction, in the last fraction and usually in every intermediate fraction. They do not at once take up their growth (at least visibly) in infusion broth mediums, nor commonly on the first agar plate seeded with the fresh filtrate. How much time is required for their development into the visible G colony form depends, in part, on the grade of candle employed, on the age and the state of the culture at the time of filtration, and on the methods of handling the filtrate.

Why the elements present in fresh filtrates from V candles more frequently show distinct colony development on the first plate on which a drop or more of the filtrate is streaked while filtrates through N and W candles more commonly require a serial passage over from two to six agar plates before colony growth can be recognized is a question of much interest. The question also arises whether the bodies that pass the filter candles are identical with the cellular elements observable microscopically in stained preparations of the G type cultures, or whether they are merely associated with the latter elements. Both of these points will be considered in the last section of this study.

In considering the nature and the results of these experiments on filtration there is one aspect that we wish again to emphasize, since it involves one essential difference between our methods and those commonly employed by other investigators. In all the studies on filtration reported in the literature, so far as we are aware, it has been the practice to filter a "normal" culture or some infectious material supposed to contain that culture, under conditions which afforded no special reason to suspect the presence of "filtrable forms," and then, to attempt the discovery of the filtrable elements in the filtrate.[14] This is far from the course that we have pursued. As indicated at the beginning of this study, we have not based our inquiry regarding the existence of a filtrable form of the Shiga bacillus on the attempt to pass through filter candles any or every culture that presented itself, since we have realized from the outset that probably the majority of "normal" laboratory cultures of this species, as of other species also, are "nonfiltrable"—to employ the term in its commonly accepted meaning. We have, on the contrary, first attempted to discover preformed in the culture, or to produce from normal S type cultures (or sometimes from the R type),

---

14. Exceptions to this statement are to be found in the recent work of Mellon and Jost and the still more recent work of Panek and Zakharoff on the filtrable form of the tubercle bacillus. Mellon used for his tests a culture the individual organisms of which were characterized by the presence of "granules." It seemed logical to expect that the next point of advance in the study of the tubercle ultravirus would be the discovery of the cultivability of the "granules" alone without the rod forms. This is what Panek and Zakharoff appear to have accomplished in developing their micrococcus and granular stage of the organism. In other words, this granular stage seems comparable with what we have termed the G type culture.

elements at such a stage of culture development that they might be able to pass the Berkefeld candles, and then, if possible, to demonstrate their filtrability. In other words, the G type cultures used in the filtration tests reported in this paper, whatever their true nature may be, were not produced by filtration. They were produced by instigating the dissociative reaction in the mother culture. After that, they were isolated, grown in pure culture and demonstrated to be filtrable. After that, it was shown that they were capable of yielding, in good time and in the presence of adequate growth opportunities, cultures that were analogous in every respect to the G type cultures originally employed for the filtration tests. And finally, as we shall demonstrate later, they were caused, by a slow and tedious process, to return to the Shiga culture of conventional type and attributes (see section on reversion).

The commonly accepted and surest criterion of the filtrability of a substance (barring imperfections in the filter and in the technic) is the discovery on one side of the filter candle or membrane, of something that was previously present and recognized on the opposite side. That is why *B. prodigiosus* or the fowl cholera bacillus is often used for testing candles and for controlling filtration procedures. If the substance or thing discovered in the filtrate was not present and recognized in the substance to be filtered, before it was filtered, the grounds for concluding that this substance actually passed the candle are not secure. It might have been present before filtration, but the evidence of this is purely circumstantial. It is in cases like this that one is inclined to seek an explanation of the apparently successful results in terms of "contaminations" that have gained access to the filtrate. If the alleged "filtrable form" of the tubercle bacillus could be produced, cultivated and identified before filtration, and could then be recognized by cultivation methods after filtration, as we have done in the case of the Shiga bacillus, it would serve to eliminate much uncertainty that many investigators believe still surrounds the filtration work with this bacterial species. As a matter of fact, Panek and Zakharoff appear to have fulfilled this requirement for the first time, as indicated in a foregoing footnote. Cooper and Petroff did not succeed in their attempt, but Mellon approached the result very closely, and Panek and Zakharoff's results will no doubt soon be confirmed. In any case, in the results which we have reported for the Shiga bacillus (and the same is true of other species studied in this laboratory) it will be observed that grounds for these criticisms do not hold. We desire to point out once more that we have not produced the filtrable form of this species by filtration.

On the other hand, it is true that the process of filtration interposed in the development of the Shiga G type culture exerts an influence in modifying this culture in at least two important respects. First, it transforms a culture that is cultivable on agar into one that, for a time at least, is noncultivable (visibly) either on agar or in broth. This apparent transformation, however, may be merely a separation of certain elements that are less readily cultivated from others that are more readily cultivated. In the second place (but perhaps only as a result of the effects mentioned), an interposed filtration has a tendency to stabilize the G form against too rapid a retransformation to, or toward, the original S culture type. It is readily noticeable that some G colonies in a strain that has not at some time in its history experienced at least one filtration show a marked tendency to become retransformed rather quickly into the "normal" culture. Submission even to one filtration through an N or W candle, changes this: it produces a G type culture that undergoes reversion only by much slower steps and after a much longer time. It is as if such a culture were much further removed from its goal. For these reasons, we are inclined to regard the G type culture, before filtration, as made up of variable culture elements, some of which stand nearer to the filtrable stage, while others have gained some headway in their progress toward the original or "normal" form. On such a variable population of cell forms, all of which differ markedly from the "normal" culture type, filtration appears to act by selecting those elements that stand nearer to the lower limits of growth, and that (one might imagine) have further to travel on their ontogenic path before they can attain the cell form of the "normal" culture—via the route of the G type. From this, one might arrive at the notion that the G type culture, as we view it in its cultivable state, is merely a stage of transition bridging the gap between the primitive, filtrable virus form and the original culture type. That the filtrable bodies (N and W filters) are not, as a rule, identical with the majority of the elements constituting the typical G culture, but are closely associated with them, and to an extent perhaps generated by them, is a point in favor of which some evidence exists, and one which will be considered in greater detail in a later part of this paper.

In view of the frequent necessity of making serial plate transfers from the first plate seeded with the filtrate, before any evidence of culture growth could be secured, and of the consequent danger of contaminations which such a method is sure to involve, we are quite ready to hear raised by our critics the objection of "contaminations" to explain the advent of the G type cultures. For this reason, even before presenting the control tests in a later section, it may be desirable, at this point, to observe the implications following such an interpretation. In the 200 or more filtrations that we have performed we have had in all instances except 2, positive assurance, by cultivation methods, of the presence of the G forms in the cultural material submitted to filtration, before any filtrations were performed. In the filtrates from these 200 or more tests we have failed in only 3 instances to discover the same culture in the filtrates, and to find it in a pure state. We have never been able to discover in our filtrates the astounding assortment of contaminations representing such a remarkable series of bacterial species reported by Frobisher (1928) following his filtration technic, whatever it may have been.[15] If the culture which grew from our filtrates, regardless of the type of candle employed, of the amount of negative pressure and of the filtration time, are to be regarded as contaminations, certain circumstances must be true: (1) The contaminations were present in practically 100 per cent of the filtrates; (2) they were represented by the same type of organism in 100 per cent of the cases, to the exclusion of any manifestly foreign bacterial species such as are common in laboratory air; (3) the contaminations, through some remarkable play of chance, were identical with the organisms present in the cultures submitted to filtration. Suspicions of this sort are difficult to entertain. But they have been entertained; and these, with other points of possible error, have been made the subject of numerous control tests to be presented in the following section.

In closing the present section it may be added that we feel justified in the conclusion that a certain form of culture found intimately associated with cultures of the Shiga dysentery bacillus at certain periods of its development has been demonstrated as readily and invariably filtrable. The chief point of issue in this study, therefore, now becomes, not

---

15. Frobisher revealed few details of his filtration methods. Unless one knows the grade of candle employed, the pressure used, whether positive or negative, the amount of pressure, the amount of substance filtered, the approximate time of filtration and the kind of filtration assembly used, it is impossible to repeat the experiments of another worker. In the meantime, both Frobisher's results and conclusions, obviously intended to offer support to the conclusions earlier arrived at from the questionable experiments of Brown and Frobisher (1927), must be regarded with much suspicion, and as possessing little significance in the furtherance of the general problems of bacterial filtration.

whether this form (the G type culture) in filtrable, but whether it is actually related biologically to the Shiga bacillus in broth cultures of which it was first produced by dissociative action. In other words: What is the actual nature and significance of the G type cultures? Do they represent a new species, hitherto undescribed, living in such intimate association with the Shiga bacillus that application of the usual methods of colony selection fail to separate them? Or do they represent one of the heretofore missing cyclostages in the life history of the Shiga species? Our belief is that the truth will be found in the last-mentioned hypothesis. On the other hand, such a conclusion finds no direct support in the experimental evidence presented in this study to the present moment. Before final conclusions can be accepted, it is necessary to examine two further aspects of this investigation—first, the important matter of the control tests, and second, the studies focused on the problem of reversion of the G type culture to the original form. These important aspects of the study will be given attention in the two following sections.

CONTROLS ON FILTRATION AND OTHER TESTS

From the data presented in the two sections just preceding we have reached the tentative conclusion that the G type culture is either itself filtrable or gives rise to filtrable elements in considerable number. For practical purposes it is sufficient to say that the G type culture is filtrable, in contrast with the S and R type cultures, which are not—at least in the sense in which the term filtrable is at present employed. The chief point at issue in the major problem, therefore, is one that from now on does not concern the question of filtration, but merely the possible relation between the G type cultures and the "normal" Shiga cultures with which our experiments were started. Is the G form a contamination from the air of the laboratories, from the hands of the workers or from the agar or broth mediums? Or does it represent a definite cyclostage in the development of the Shiga species? With these possibilities open, it seems desirable at this point to record certain further details regarding the technic employed; also to mention some of the tests used for purposes of control.

CONTROLS ON CULTURE MEDIUMS AND ON TECHNIC OF PLATING

All the broth mediums were prepared in the laboratory from infusion stock. This was also true of the agar, with the exception of a portion of the litmus lactose agar, which was a "Difco" product. This, however, was used only at a time in the investigation subsequent to the first production of the G forms in broth and their cultivation on plain infusion agar and on laboratory-prepared litmus lactose agar. The infusion mediums were sterilized at 15 pounds (6.8 Kg.) pressure (121 C.) for from twenty to thirty minutes, depending on whether they were contained in culture tubes or in the 300 cc. storage bottles. The litmus lactose agar was usually sterilized for thirty minutes at 110 C., and in some cases for thirty minutes at 121 C. In the latter instances, some of the sugar was probably affected, but the aim was to be sure that no organisms survived heating, and the changes that may have occurred in the medium were certainly of no disadvantage in the work. Indeed, there was no indication that our Shiga cultures fermented lactose. This aspect of the study received special consideration in Miss Raven's work with the G forms of *B. enteritidis*.

In view of the somewhat greater danger in using litmus lactose agar sterilized at lower temperatures than those employed for the other mediums, and in view of the important part played by litmus lactose agar in our investigation, it is desirable to call attention to a point of special significance. Recognizing this possible difficulty, we made it a rule not to employ litmus lactose agar plates for the demonstration of the G type colonies until these, in the case of any given strain, had first been demonstrated, isolated and studied, in part, on plain infusion agar plates and in infusion broth. This procedure would seem to rule out the possibility that the G forms could have originated in the frequently less perfectly heated litmus lactose agar.

The mediums were "tested for sterility" by the usual method of incubating for from eighteen to twenty-four hours or longer (in some cases, for several months) at 37 C. Since, however, we have come to the conclusion that such incubations for sterility may be inadequate, under certain conditions of growth, for revealing the presence of all living elements, we subjected samples of our broth mediums to the same sort of tests as those used in the propagation of the G type cultures from fresh filtrates of G type cultures, that is, the serial agar plate method first presented by Hauduroy (1927, a).

Blood agar plates and ascitic fluid agar plates were occasionally employed for purposes of comparison, but never when these were not controlled by plain infusion agar plates. Moreover, as mentioned earlier, these mediums containing blood, ascitic fluid or peritoneal exudates were never employed for any crucial test, and none of our conclusions are dependent in any way on the results obtained from the use of such mediums. Indeed, we are inclined to be suspicious of such mediums when used for any study of microbic dissociation in which they are not positively demanded by the conditions of the experiment.

Obvious contaminations occasionally arose on the plates after exposure to the air, but they seldom occurred before the plates were opened for the purpose of microscopic examination, for the taking of impression smears of colonies or for culture samplings. When contaminations were observed, they embraced the usual range of laboratory contaminants. Plates accidentally contaminated and those contaminated intentionally by exposure to laboratory air never revealed, except in one instance, the surface films or pinpoint colonies characteristic of the G type cultures.

In view of the possibility, however, that "invisible colonies" arising from the air or from the hands might be present on the plates, but undetected, the bare agar surfaces lying between the contaminating colonies on such intentionally exposed plates were washed up with sterile broth, and this broth was transferred to a second plate and so on, in series, according to the technic employed in recovering cultures of the G type from the broth filtrates. The plates remained free from colonies in all cases, except one. In this case there appeared a fine film of apparent growth which bore some resemblance to the G culture. On repeated transfer, this film gave rise to minute, whitish colonies having a diameter of from 0.3 to 0.4 mm. Microscopic examination revealed large micrococci having a diameter of from 1.5 to 2 microns. With continued transfer, these colonies became larger and finally developed into a yellowish, glistening growth resembling that of some of the chromogenic air cocci. It seemed possible that the plate of origin became contaminated with a slow-growing air form or with the G form of such a culture, which eventually reverted to the original type. It would be surprising if the G forms of other species were not present in nature, as in air, soil, dust and water. In fresh sewage, for example, it is practically impossible to escape the filtrable forms of several bacterial species which easily pass the V and N candles. The observations of Hauduroy (1926, a) and of Fejgin (1925) in relation to the filtrable forms of B. typhosus and other species lend credence to this view.

To the foregoing comment it may be added that in the considerable number of years of laboratory experience which one of us has enjoyed he has never previously observed in mass, on plates of any sort, contaminating colonies that bore a resemblance to the G type colonies here described. The contamination hypothesis is difficult to maintain, and particularly so in view of the fact that, as we shall undertake to show later, we have been able to bring about a retransformation of the G type cultures to the original forms from which they first arose. On the other hand, we believe that in future work in this field it will be necessary to guard carefully against air contaminations with adventitious organisms that also may be found in the G state, and that may, at a certain stage of their development, give slight, or no, evidence of growth on plates or in broth mediums.

Since our early results seemed to throw some discredit on the commonly employed methods of testing for the sterility of culture mediums in general, it seemed desirable to perform special tests on samples of the sterilized, and presumably sterile, broth used in our work. Accordingly, broth removed from stock tubes was spread, in amounts of from 5 to 10 drops, on presumably sterile agar plates. The first plate was incubated for forty-eight hours. Fresh broth was then added, spread, collected and transferred to a second plate. This procedure was carried through a series of six or more plates. The plates often manifested, after incubation, the presence of a faint film, somewhat resembling those seen on plates which eventually revealed G colonies. Repeated transfers of this material, however, never revealed G type colonies on later plates.

Controls on Pancreatin Broth.—The controls on the medium containing Squibb's pancreatin demand special consideration, in view of the manner in which it was sterilized and the results that were obtained from its use. The common method was merely to heat at from 60 to 62 C. for one hour in a water bath. The relatively large number of instances in which this medium gave rise to the G type cultures might suggest the presence, in the raw pancreatin, of organisms that could be confused with the G cultures submitted to filtration. The 5 per cent pancreatin broth tubes were incubated at 37 C. for periods between two days and four months. On casual examination no sediment or other sign of growth was observed. But it has been pointed out that the sterile appearance of a broth tube or of an agar plate may reveal little information regarding its actual sterility. For this reason, the pancreatin broth tubes were subjected to a more exacting test. An example

Five drops of the presumably sterile pancreatin broth were streaked over the surface of each of sterile litmus lactose agar plates (previously tested), and these were incubated for forty-eight hours or more. If no film or other evidence of culture growth appeared within this time, several drops of the broth washings were transferred from the first plate to a second, then from the second to a third and so on through a series of six plates. At no time in the course of such transfers did we observe any cultures or colonies resembling the G type. To the time of exposure to the air for the purposes of examination with microscope or hand lens, they remained sterile so far as we could ascertain.

In addition to the tests mentioned, others were performed by filtering the presumably sterile pancreatin broth through a sterile candle, and then effecting serial plate transfers, exactly as in the chief experiments. Such plates, six in series, remained free from any evidence of the G colonies or foreign contaminants.

In another test, a sample of pancreatin extract was heated at 60 C. for thirty minutes, then filtered through an old Berkefeld N candle under less than 200 mm. of negative pressure. The filtrate was distributed into several tubes, which were incubated for five days, then kept at room temperature for more than four months. All remained sterile. Plates of litmus lactose agar were seeded with liberal amounts of the filtrate and incubated for several days. They remained sterile. Serial plates were then employed to the extent of five, at twenty-four hour intervals. All remained sterile.

In order to ascertain whether any filtrable forms of other bacterial species were present in the pancreatin solutions that had not been heated, the following test was performed. A 5 per cent pancreatin solution in sterile broth was filtered through an old Berkefeld candle and dispensed into several tubes that were then incubated at 37 C. for four days, after which they appeared sterile. They were then kept at room temperature for seven days. At this time, they revealed a heavy, whitish, mucoid sediment. Direct plating on agar showed medium-sized, whitish, glistening colonies made up of large cocci. We tentatively concluded that this organism was present as a contaminant in the pancreatin, and that a filtrable stage might have passed the candle. It was manifestly eliminated from the pancreatin by heating. In any case, it bore no relation to the G type cultures.

Finally, it may be added that, even though one may desire to regard the G type cultures as contaminations arising from the pancreatin, all work employing this medium might be eliminated from consideration, and our conclusions would be the same. Pancreatin broth was only one of the means by which the G forms were produced.

CONTROLS ON FILTER CANDLES AND ON HANDLING OF
FILTRATES

As pointed out earlier, the possibility of contaminations of the filter candles as a source of the G forms has no significance in this work, since the G type cultures were produced, isolated in pure lines and studied before any of them were submitted to filtration, and we were able to demonstrate that the cultures eventually obtained from the filtrates were identical with those observed before filtration.

Used candles were boiled, washed by direct and reverse flow of distilled water, and sterilized as mentioned earlier. Through such candles was passed, under 200 mm. of negative pressure, 25 cc. of sterile broth such as that used in the investigation. Of the collected filtrate, 10 drops were spread over the surface of several sterile agar plates, and then, after incubation (no growth appearing), the broth washings transferred serially, at twenty-four hour intervals, over other agar plates, exactly as in the case of the G culture filtrates. These tests, involving sterile candles and sterile broth, were performed four times, once with a V candle, twice with an N candle and once with a W candle. The serial plate to plate transfers were carried over a total of eight plates. G colonies never appeared.

It might be held by some that the danger of contamination was greater in the fractional filtrations than in the single-volume filtrations. The following tests were therefore performed with presumably sterile broth. First, a single volume was filtered through a Berkefeld candle that had been used once. The candle, after removal of excess of water, was found to hold 640 mm. of negative pressure. After being again assembled by Mudd's method, the candle was washed, freshly sterilized at 130 C. for thirty minutes and then set up with paraffin seals over the rubber stopper and metal connections. Through this candle was passed under 120 mm. of negative pressure, 125 cc. of 7.4 infusion broth (more than three times the amount ordinarily employed). The filtrate was taken in five fractions, the pressure being released for each fraction. Each fractional filtration represented about thirty seconds in time, and from 5 to 10 cc. in amount. All the fractions were then placed in sterile tubes and incubated at 37 C. for thirty days. At the end of this time, all proved sterile, both by simple observation and by the serial agar plate method. At the end of four months they were still sterile. The first and second fractions gave a slight sediment. In neither case, however, was this cultivable, and microscopic examination revealed the fact that it was composed of fine particles of filter substance.

In a second series of tests, a single large volume of filtrate, similarly obtained, was divided into seven fractions, the total volume being about 40 cc. The filtration time was about two minutes through a Berkefeld N candle under 160 mm. of negative pressure. The fractions were taken into sterile tubes, which were at once sealed and then incubated at 37 C. for thirty days. One tube was cracked and discarded; the others remained sterile by all methods of test.

The results of these and other tests performed as controls on the filtration technic leave us with strict confidence in the methods of filtration employed in our work; also in our methods of handling the fresh filtrates. With good candles and perfect technic, contaminations in Berkefeld filtrates should be rare.

TESTS WITH HEATED FILTRATES

If the filtrates of the G type cultures contained living elements capable of regeneration into, first, the G type cultures, but eventually into the "normal" form of the Shiga bacillus, it seemed probable that a filtrate that had been heated for one hour at 60 C. would not give rise to the G type culture, either on the first plate or on any subsequent plate in the series on litmus lactose agar. In order to test this point, a fresh filtrate of G type culture, strain I, was used. The culture employed represented the twenty-third broth transfer in a series in which tubes 1, 3, 5, 7, 9, 11, 13, 15, 17 and 19 had all given positive results on filtration. The culture in tube 23 was divided into two samples, one for filtration through a V candle, the other through an N. The V filtrate, unheated, gave positive results on the fifth or sixth serial plate, the lateness here being an exception to the general rule. The N filtrate, unheated, gave results on the third or fourth serial plate. Samples of the V and N filtrates, heated in the water bath for one hour at 60 C. and then passed over serial plates, gave no sign of the G colonies in a series of eight plates. These results were in harmony with the outcome of other tests in which it was demonstrated to be impossible to obtain G type cultures either from heated filtrates or from the unheated filtrates of G cultures that had been heated before filtration occurred.

To the foregoing observations it may be added that, in order to afford a control on the controls and to demonstrate that the G stock culture employed was still viable at the end of the tests, a further transfer was made through tube 24 to tube 25 in the series. Culture 25 was filtered in three fractions, through a V, an N and a W candle. The V filtrate gave colonies on the fourth serial plate, while the N and W filtrates both gave the first G growth on the fifth serial plate.

CONTROL OBSERVATIONS ON THE "FILMS" OCCURRING ON THE PLATES SEEDED WITH THE G CULTURE FILTRATES

The "films" that have been mentioned earlier seldom covered the entire plate. They occurred in streaks or small patches and never transgressed the areas seeded. The question naturally arose whether this filmlike substance which was usually present on the early plates of a series before the appearance of the G colonies, and which immediately preceded the G colonies, represented extraneous material from the broth or from the agar surface, or whether it was a form of culture development. In studying this point, filtrates that were known to have given the G colonies by the usual method of plate to plate transfer were sealed in glass ampules and heated in the water bath for one hour, sometimes at 60 C., and sometimes at 62 C. They were then opened and spread over sterile agar plates by the usual procedure and in the usual amounts. Surface films, somewhat resembling those produced by the unheated filtrates, often developed. The surface of such plates was then washed up with a few drops of sterile broth, and the washings were spread over fresh plates and so on in series. This attempted cultivation of living elements in the form of the G type culture, through a series of from six to eight plates, never gave positive results. Moreover, the examination of the films by impression smears stained by the Giemsa and Romanowsky methods, although often revealing minute, coccus-like bodies, and occasionally rodlike structures, presumably arising from the stains themselves, never revealed the aggregations of minute cocci, the long filaments, the peculiar sheathlike formations and the streptococcus and staphylococcus groupings that were seen in films on plates treated with unheated filtrates.

We also prepared impression films from sterile agar plates. Even under these conditions, we found that this procedure may reveal a considerable number of bacteria-like bodies that are likely to be confusing to one who has not made a study of them, and might easily lead to erroneous interpretations. It should be made clear, however, that the difference between the impression smears made from plain sterile agar plates or from sterile plates streaked with heated or with unheated filtrates or with sterile broth were sufficiently clear to permit of no more than slight confusion. In none, except those streaked with the unheated filtrates, could be observed the typical pictures that have been described. We believe that our observations in the microscopic examinations of the surface films have been adequately controlled. We are therefore inclined to the view that at least some of these filmlike appearances are concerned with the development of living elements of some sort, which precede the appearance of the typical G type colonies and culture.

CONTROL OBSERVATIONS ON THE "GROWTH SEDIMENTS" [16]

Since minute traces of sediment often appear in tubes of sterile broth after sterilization, it was a matter of importance to demonstrate that the sediments described earlier represented a form of growth of the filtrable elements, even though the supernatant fluid remained clear. It was suggested that the mere procedure of inoculating a tube of sterile broth with a sterile platinum wire might provoke the appearance of similar sediments. This was tested. The results showed that a sediment was sometimes produced in this manner in a perfectly clear medium. These artificial sediments, however, differed from the "growth sediments" in that they were less voluminous and either less compact or less viscous, and that they did not increase with the lapse of time. Moreover, when these sediments were examined in stained preparations, although minute, granular bodies, staining bluish or bluish purple with Giemsa, were often seen, the picture as a whole in no way resembled those obtained from the "growth sediments." Moreover, such artificially produced sediments never gave rise to G colonies on serial plate transfers. The same was true of the sediments artificially produced in sterile broth by the addition of alcohol or mercuric chloride.

Another source of possible error in connection with the sediments was found in the interpretation of deposits in the mediums, arising from loosened particles coming from the freshly baked candles. To demonstrate this point, the following test was performed.

A Berkefeld candle, used once, was set up in assembly, impregnated with water and the surplus removed. The candle was then found to

16. Many of these control measures were suggested to us by Dr. Novy.

hold about 640 mm. of negative pressure. It was then dried and baked in the oven, cooled, again set up in assembly and the assembly sterilized under steam pressure at 130 C. for thirty minutes. Through the sterile candle was then passed a quantity of sterile broth, which was collected in five fractions. The filtration time for each fraction was about thirty seconds. Each fraction was sealed in a glass tube and incubated. In the first two tubes, a definite precipitate was formed, greater in the first. The remaining fractions remained without sign of growth or of sediment. Microscopic examination of the sediments of the first two fractions revealed a type of débris showing no similarity to the growth sediments. The material was composed of a fine substance manifestly derived from the filter candle. Further, all attempts to propagate this sedimentary matter resulted in failure. We cannot therefore agree with those investigators who have stated that filter sediments may be mistaken, microscopically, for such growth sediments as have been depicted by Hauduroy (1927, a), which have been studied by ourselves, and which are manifestly due to the presence of living, propagating elements in the broth medium. Such conclusions can arise only in the minds of those who have not observed personally either one or the other of these phenomena.

SUMMARY OF SECTION AND CONCLUSIONS

Up to the beginning of the present section it had been demonstrated that a form of culture (the G type) found associated with cultures of the Shiga dysentery bacillus at certain times in their development passed Berkefeld candles of all grades readily and was eventually recoverable by cultivation methods in practically 100 per cent of cases. The question therefore arose whether the G type culture could be regarded as a contamination the nature and origin of which might be revealed by recourse to control tests relating to the various procedures entering into the laboratory experiments. In answer to this question, the results presented in the present section have shown that we have not been able to observe or to produce, by any method involving the use of sterile reagents, mediums or filters, either colony forms or cell types bearing resemblance to the G type cultures reported in this study. These results strongly suggest, but do not necessarily prove, that the G type cultures are not contaminations of the original culture or contaminations arising in the filtrates, but are a stage in the development of the Shiga bacillus. The final and incontrovertible proof that the G forms are developmental units in the Shiga species must comprise the demonstration that these bodies are able to become retransformed into the original and "normal" culture type. It will be the aim of the following section to present this evidence.

REVERSION OF G TYPE CULTURE TO ORIGINAL FORM

Up to the beginning of the present section the chief points that we have attempted to make clear are the following: 1. There has been found present in "normal" Shiga cultures, under the influence of certain incitants to the dissociative reaction, or sometimes in unmodified mediums, a peculiar form of culture which we have termed the G type and which differs markedly from the parent form, as well as from recognized cyclostages of the Shiga species. 2. This G type culture is readily and invariably filtrable through Berkefeld candles. 3. From the filtrates the G forms can easily be obtained if the proper technic is employed. The question now confronting us is: What is the nature and significance of this new culture form? Since, in the preceding section, it was clearly indicated that the G cultures were not a contamination arising through any of the laboratory procedures, the only considerations left are the following: 1. The G cultures represent a contamination that was present from the beginning in the original stocks, but that revealed itself only under certain conditions; perhaps a new bacterial species, hitherto unrecognized, which lives in such close association with the Shiga bacilli that separation of the two by common plating methods is impossible. 2. It is a distinct filtrable stage in the cyclogeny of the Shiga species.

Except for one circumstance, to be mentioned presently, it is not an easy task to rule out of consideration the first possibility. The circumstance of the more or less sudden appearance of the G forms in the assumedly pure Shiga stocks does not itself prove that there exists a biologic relationship between them, and the same may be said regarding analogous G types found by ourselves or by others to be associated with *B. typhosus*, *B. paratyphosus* and several other species. Moreover, the great similarity, morphologically and culturally, between the G type Shiga cultures and the G forms from these other bacterial species might seem to imply that one, single contaminant, present in our broth, agar or laboratory air, lay at the basis of all of these observations on the origin of the G forms. As noted earlier, we discount entirely the rôle of filtration in giving rise to these cultures, for the reason that they were produced and cultivated before any filtrations were attempted.

Regarding the interpretation of the G forms as a new bacterial species living in association with the Shiga bacillus—this conception, though used much of late to explain the coincidental appearance of certain filtrable virus-like agents with microscopically visible culture forms (as in the case of *Streptococcus scarlatinae*, for example) has little to recommend it. This phase of the subject will be considered more at length in the following section, but it may be said in passing that if such an argument is used to reflect discredit on the possible biologic relationships between two associated elements (one filtrable and, with difficulty, cultivable; the other nonfiltrable and easily cultivable), the argument merely results in demonstrating that there does exist a form of cultivable growth that is readily filtrable, whatever the species may be. Such arguments can easily lead to useless equivocation.

But, we believe, there is little need of our attempting to combat further the notions of "contamination" or of "association," for the same body of evidence refutes them both; that is, the demonstration that the G type cultures, in the course of continued propagation under suitable conditions, undergo a retransformation into cultures that, so far as we have been able to ascertain, are identical with the original form of the Shiga bacillus from which they were derived. It is therefore the aim of the present section to give evidence supporting this fact.

The "reversions" to be mentioned occurred in seven of the G strains studied. Some others, kept for one to two years in sealed ampules, have thus far retained their characteristics. At present these appear to have become established in the G form even more securely than the S and R forms can be established in their respective culture types. Still other G strains have not been studied especially with a view toward enforcing their retransformation. In two cases the retransformation was to the R, instead of to the S, type Shiga.

Although it often happens that some cultures that approximate the G form return rather rapidly to the original state when propagated on solid mediums, the evidence to be presented involves the reversion of strict G types and is of two sorts. One group of data deals with the continued propagation, from tube to tube, of the mass culture in which the G forms first arose, but without the isolation and purification of the G culture. This evidence is not entirely trustworthy, since it reveals merely certain sequences of culture growth without supplying the definite proof that one form actually developed into another. The second group of data deals with the recognized transformation of pure line strains of the G type into cultures that appear to be identical with the cultures of origin, that is, the "normal" Shiga stocks. It is on this second category of evidence that our conclusions regarding the fact of reversion are primarily based. At the same time, we believe that these results lend greater significance to the somewhat questionable reversions in the first group.

REVERSION FROM G TO S FORM IN CULTURES UNDERGOING MASS CULTIVATION

Bearing on this point an inspection of the tables presented in the section on the origin of the G cultures indicates several examples of the recovery of the S or the SR forms after the agar plates (one or more) had given the appearance of sterility or had revealed G colonies only. This point is brought out in the following instances: table 1, strain, Ia, transfers 16, 17, 21 and 22; table 2, strain IIa, transfers 14 to 38; table 3, strain III, transfers 12 to 14 and 27 and 28; table 4, strain IV, transfers 11 and 12, and table 5, strain V, transfers 8 and 9.

In these cases it appears that the S or R colonies reappeared after a longer or shorter absence. Perhaps the chief reason for doubting these cases as examples of reversion is the fact that the time allowed for the change was often very short as judged by the cases next to be presented. We therefore prefer to leave open the interpretation of these results.

REVERSION FROM G TO S OR R FORM IN PURE LINE G STRAINS

The first pure line G strain to manifest complete reversion was a subculture of strain I, the origin of which from an R type Shiga culture has already been indicated in the section on origin of the G cultures. In this case, the reversion to the R type took place very slowly, being accomplished by about sixty serial transfers at twenty-four hour intervals through plain infusion broth. Frequently, for a week or more of daily passages, it was impossible to detect any change in colony form. Over the entire period, however, it was observable that there occurred a gradual increase in the size of the G colonies from 0.2 mm. or less to about 2 mm. (the average size of the S colonies) or to from 3 to 5 mm. (the average size of the R colonies). This slow change was accompanied by a loss of the delicate structure and of translucency and by an increase in opacity. It may not be accurate to say that the reversion was completed at the sixtieth passage. It might have been complete at the fifty-seventh or the sixty-second. When the changes occur so gradually it is difficult to draw the line at the exact moment of attainment of the original culture characters. We therefore say that it was at about the sixtieth passage when the G form seemed to have reverted completely and to have taken on the biochemical, and at least some of the serologic, characters of the original type. It had also again become sensitive to a mixed Shiga bacteriophage.

Strain II of the pure line G type culture, the origin of which from an S type has also been depicted in the section on origin of the G cultures, reverted to a typical Shiga S type culture after about fifty-three daily passages through plain infusion broth. This retransformation was also very gradual, but eventually the culture came to resemble the original form. It also became sensitive to the Shiga bacteriophage. In the G state, all cultures were found to be resistant.

Strain III of the pure line G type culture, the origin of which under the influence of pancreatin has already been described, reverted to a typical S type Shiga culture in a somewhat different manner. After a few early passages, the G stock culture was sealed in a glass ampule and stored at room temperature. This was opened after a period of about four months, and the culture characteristics were found to be apparently unchanged. The same minute colonies appeared on the agar plates. From this time on, over a period of about four months, the culture was submitted to broth transfers at intervals of about two weeks. At many of the transfers the cultures were plated and the largest colony selected for the next broth transfer in the series. By this selective method, the colonies in about four months time had nearly attained the diameter of the typical S type Shiga colonies (2 mm.) and resembled them in other respects. After five months, no differences could be noted. In addition, the culture was now agglutinated in Shiga S type immune serum. It resembled *B. dysenteriae* Shiga biochemically, was toxic for rabbits, and had again become susceptible to the Shiga bacteriophage, to which it was earlier resistant.[17]

In all of these tests of the identity of the "reverted" strains, we are inclined to lay considerable stress on the susceptibility to the Shiga

17. Methods that failed to cause a reversion of G strain III to the original culture type were as follows: (1) sixteen serial transfers at forty-eight hour intervals across litmus lactose agar plates; (2) ten serial transfers at somewhat irregular intervals through 10 per cent anti-S Shiga immune serum broth.

bacteriophage, since this principle was not directly active on either *B. coli* or *B. typhosus,* or on other species less closely related to the dysentery bacillus.

In connection with the question of stability of the G forms of culture, we have noted one other point of interest. Although some G strains freshly isolated without filtration remain stable over long periods on further propagation, and although rarely some strains freshly developed from V filtrates may revert fairly quickly (i. e., in from five to eight transfers) to the large S colony type, the following relations, in general, appear to hold true. Strains at some time in the history of which a filtration has not been interposed revert quickly; while strains in the previous history of which a filtration has been interposed change much more slowly. In other words, the selection of certain elements determined by the filtration process seems to exert a stabilizing influence on the G forms. If a G culture has shown a tendency to "grow up," but has not gone too far in this direction, it can be brought back to the minute G form by another filtration. Filtration at intervals may therefore be employed to hold a G culture "true to type." In some strains, however, this measure does not seem to be demanded, since, in broth cultures at least, the G forms show little tendency to advance toward the large colony form. This is especially true if they are stored in sealed ampules, as will be mentioned later (section on viability of the G forms).

Another matter dealing with the rapidity of reversion concerns the proper selection of the G colony used as the basis of the prospective G type culture. It frequently happens that an agar plate at the end of twenty-four hours' growth yields a considerable number of very small colonies, but that subculture from such colonies may not yield a typical G stock. There are various reasons why Shiga colonies may be small at this stage of development. They may be crowded; the agar may be thin over the plate; it may be too dry, or the medium may not be of the proper reaction. Cultivation from such colonies will almost invariably give, on the next plate, colonies of greater size. The essential criteria of the presence of a G colony are that it develops slowly on the plate, that it is minute even though the colonies are well spaced, and that it remains minute even after from six to ten days of incubation. For this reason, for the purpose of isolating a G strain, it is preferable to select a colony that is well spaced from its neighbors, and to select one that remains minute (0.05 to 0.2 mm.) after a period of ninety-six hours' growth on a favorable medium sufficiently deep in the plate. The most stable G cultures arise in broth tubes from which all other cell elements (R and S forms) have disappeared. Selecting G colonies from a plate on which there are present all possible gradations in size from minute to large ·G forms, and even some that might be regarded as beyond the G stage, is seldom successful in establishing a stable G culture. Merely the gross appearance of such a plate is sufficient evidence that the G forms, in this case, are not young forms, but have been caught in the course of their reversion to the mother type. But, as already noted, the most stable G type culture of all is one derived from the fresh filtrates of a young G broth culture.

Our final point relating to the reversion remains to be mentioned. It is the extremely slow and gradual process of retransformation compared with the relatively sudden generation of the G colonies at the beginning, when they first arise from the S or the R cells through dissociative action. While their first origin is what might be termed "explosive," their return is likely to be a matter of weeks or months. In the intermediate steps of this journey they would not be regarded as members of the Shiga species. What the gamut of biochemical and serologic characteristics of these intermediate stages may be, we do not know. Nor have we, to the present moment, ascertained at what point in the gradual change from the G to the mother culture filtrability is lost, or at what point virulence, toxicity and susceptibility to the bacteriophage are regained.

SUMMARY OF SECTION AND CONCLUSIONS

The data presented in this section conclusively demonstrate that, just as the filtrable bodies (the Shiga virus) become transformed into the elements of the G type culture, this culture, in turn, becomes transformed into the "normal" culture. The change from the filtrable elements into the G forms is obviously much less rapid than the change from the normal forms into the filtrable elements. The transformation from the G forms into the mature culture type is even slower in its course, as has been clearly seen. And this is especially true if the G culture has recently arisen from filtrates. We have spoken of the retransformation of the G forms into the cells of ordinary type as a "reversion." It would perhaps be more accurate to regard it as a "progression"; for we must look on the G forms as still primitive growth elements lying close to the virus forms. They are the stages representing the "physiologic youth" of the culture. The term "regenerative bodies," as employed by Löhnis (1921), might be employed for the cultivable elements of the youngest G type cultures.

Taking into consideration these facts, with other data presented in earlier sections, the unprejudiced reader, it seems to us, can arrive at only one conclusion. The G forms are not subtle contaminations of the original cultures; they are not organisms of a foreign species living in close association with these cultures. In reality, the Shiga species is a composite of heterogeneous cell forms, among which the S and R are the most common. In addition, there are the virus form and the elements of the G type cultures. Into the latter the filtrable forms develop, and out of them again very slowly emerges the "normal" Shiga type. If any fungus producing conidia or any higher form producing seed may be said to possess a cycle of development, the same is true of B. dysenteriae Shiga. This process of passing into and out of the filtrable state appears to be the physiologic equivalent of the generation of seed forms by higher micro-organisms and of the development from the seed form to the adult. It has nothing in common with the so-called "youth-to-age" variations of Henrici and others, which chiefly concern variations among multiplying bacteria rather than among reproducing bacteria. It is rhythmical in occurrence, repeating itself at intervals in the life of the culture. The process may be hastened or intensified by establishing certain physical or chemical conditions of growth. Some of these conditions might be regarded as unfavorable; but, favorable or unfavorable, the process is bound to occur in some measure, for it rests on the most fundamental law governing the behavior of living stuff—the demand for assimilation, growth and reproduction.[18]

VIABILITY OF G FORMS AND OF ASSOCIATED FILTRABLE ELEMENTS IN SEALED AMPULES

What has been said in the preceding sections regarding the development of the G forms in or from filtrates of G cultures has concerned largely observations on fairly fresh cultures and fresh filtrates. It now remains to consider the behavior and viability of the G forms and of the associated virus forms stored in glass ampules for a period of from

18. Here it will be apparent that we should perhaps carefully distinguish between true bacterial reproduction and vegetative cell multiplication, which cannot be regarded, among the bacteria at least, as a mode of reproduction in the strict sense.

several months up to more than two years. Some of the observations now to be reported concern the behavior of the G forms present in G cultures that were sealed after from eighteen to twenty-four hours of growth. Others deal with the behavior of the G or virus forms present in Berkefeld filtrates of G cultures sealed within a few minutes after their recovery from the filter candles.

In both cases the storage was mainly at room temperature, although some of the ampules were kept for a few months either at 37 C. or at cold room temperature (from 6 to 8 C.). Darkness prevailed at all times, but we are not sure that this was of any advantage. The volumes of the sealed cultures or of the sealed filtrates varied from 2 to 6 cc. The volume of enclosed air was, on the average, one-half to one-third the volume of the medium. The ratio may actually be regarded as somewhat greater than this by reason of the heating that the necks of the tubes received during the sealing process, which undoubtedly created a reduced air tension in the ampules. We shall consider the observations on the sealed cultures and on the sealed filtrates under separate headings.

VIABILITY OF G FORMS IN SEALED CULTURES

The G type cultures destined for sealing were grown for from eighteen to twenty-four hours in beef infusion broth having a reaction of p_H 7.6. At the moment of sealing in the blast flame the growth in most cases was scant, yielding only a faint clouding of the medium. In some of the tubes it was impossible to detect any evidence of culture development at this stage.

In all, sixty-five ampules of G type culture were prepared and sealed. These represented five different G strains and fifteen different lots. In addition there were sealed two ampules of a G type streptococcus obtained from a case of erysipelas. This G culture had earlier been obtained from a filtrate. The cultures were sealed on various dates between May 12 and Aug. 1, 1928. Set away, none were examined until July 1, 1930, after a period of from about one year and eleven months to two years and two months. Cultural examinations were made first on July 3, 1930, and were continued at intervals through this month.

When after about two years the sealed ampules were examined macroscopically for signs of growth, evidences were not entirely lacking in any tube. About one quarter of the tubes showed faint turbidity or haziness, and a few revealed granulations on the tube walls. All, however, showed a larger or smaller amount of sediment on the bottom of the tube. In some cases, this was composed of finely granular material, which could be raised from the bottom by twirling the tube. In other cases, the sediment was purely viscous, arising in a swirl. In the majority of cases, however, the sediment was a mixture of granular and viscous material, the granulations appearing embedded in the viscous substance and rising with it when disturbed. On standing for a few hours, it again fell to the bottom of the tube, leaving the supernatant fluid clear. In general, it was observed that the amount of sediment

TABLE 18.—*Results of Examination of Eleven G Type Cultures Sealed in Ampules for About Two Years*

| Tube* | Date of Sealing | Appearance of Tube | Results of Cultivation† | |
			Agar Plate	Broth
I-28A-a	6/14/28	Clear; viscous granular sediment................	+	+
I-30-a	8/ 1/28	Clear; viscous sediment...........................	—	—
IV-II-14a	8/ 1/28	Clear; viscous granular sediment................	—	—
IV-IV-14a	8/ 1/28	Cloudy +; viscous granular sediment...........	+	—
I-34A-b	8/ 1/28	Cloudy +; viscous sediment......................	+	+
I-28B-a	6/14/28	Hazy; viscous granular sediment................	+	+
I/28A-b	6/14/28	Cloudy ++; viscous granular sediment..........	+	+
III-35A-c	8/ 1/28	Hazy; viscous sediment...........................	+	+
IV-II-15a	7/25/28	Cloudy +; viscous granular sediment...........	—	...
IV-IV-14b	6/14/28	Cloudy +; viscous granular sediment...........	—	...
V-7-a	7/29/28	Cloudy +; viscous granular sediment...........	—	—

* The first five ampules were opened July 3, 1930; last six ampules, July 9, 1930.
† Cultivations were on agar plates and in broth: 2 drops of sediment to plates; 1 drop to broth

present in the sealed ampules of culture was greater than in the sealed ampules of filtrate. In character, however, the sediments were much the same. The degree and frequency of clouding were also greater in the former.

The examination of the sealed cultures culturally involved withdrawing about 6 drops of the bottom sediment by means of a sterile, capillary pipet and making transfers as follows: (1) 1 drop (about 0.03 cc.) to a tube of plain infusion broth having a reaction of about p_H 7.6; (2) 2 drops to a previously poured, cooled and incubated plate of litmus lactose agar, and (3) 2 drops to a similarly prepared plate of plain beef infusion agar. The balance was used for spreading on two prepared glass slides for staining. The most important features of the first eleven sealed cultures examined in the first series are presented in table 18.

From table 18 it appears that 50 per cent of the old G cultures sealed in ampules for about two years yielded the G forms on direct cultivation of the sediments on agar plates. These results were obtained on the first plate seeded. If the serial agar plate method had been applied to the negative tubes, it is probable that some of these also would have given positive results.

As also shown in table 18, the plain broth mediums seeded with the sediments showed growth in all cases in which the plates gave G colonies, except one. Cultivation of the positive broth tubes to agar plates revealed G colonies in every instance.

Comparing these results with those obtained from the sealed filtrates (table 19), one may note that recovery of the G forms occurred in twice as many cases in the sealed culture series. Moreover, it was only in the ampules of sealed culture that the seedings of the sediments to broth gave positive results.

VIABILITY OF VIRUS FORMS IN SEALED FILTRATES

The filtrates studied in this respect were derived from filtrations of Shiga G type cultures through V, N and W candles, performed according to the technic already presented. At the time of sealing, the filtrates were water clear and, in most cases, remained so for several weeks at least. Altogether fifty-two samples of filtrates, representing a total of twenty-one different filtrations through as many candles, were sealed in glass. Of these fifty-two samples, thirty-four were from N, thirteen from V, and five from W, candles. Two additional streptococcus G filtrates were from N candles. In two cases the filtrates were heated directly after sealing and these served as partial controls on the others.

The filtrations were performed in 1928 and extended from March 12 to May 28. The examinations were therefore made, for the most part, on samples that had been sealed for a period of from two to two and one-half years. Other samples have been retained for examinations in later times.

Most of the filtrates were stored at room temperature (from 19 to 20 C.), but some were placed at 37 C. and others at cold room temperature (6 to 10 C.). The two latter groups, however, were transferred to room temperature late in July, 1928.

The tubes were examined for appearances of growth at rather irregular intervals. It is known that the majority remained clear for from one to several weeks, but we do not have a definite record of the time when the first sign of culture development appeared. We know, however, that on Aug. 2, 1928, all but two (sealed on March 14 and May 6, respectively) showed some evidence of growth. The growth was described at the time as being "hazy" or faintly granular. At that time no differences were observed that could be correlated with the grade of filter candle employed, with the time of sealing or with the temperature of storage.

When the ampules were examined again on May 1, 1930, only slight changes could be detected. The amount of precipitate in some of the tubes had slightly increased. One of the two tubes that were clear on Aug. 2, 1928, was still clear, while the other was hazy. An ampule containing a streptococcus G filtrate, sealed on May 18, 1928, showed on May 1, 1930, a granular sediment similar to those of the Shiga filtrates.

It might easily be assumed that the appearance of the tubes showing granular or viscogranular precipitates or haziness was unsatisfactory evidence of the presence of living elements in the stored filtrates. Accordingly, some of the sealed ampules were opened and cultivations made to broth and to agar plates. The first attempts at cultivation were made on March 29, 1930. At this time ampule 19A of Shiga G strain III, which had been filtered through an N candle and sealed on May 10, 1928, was opened, and 10 drops (about 0.3 cc.) were seeded to beef infusion broth of p_H 7.6, to a litmus lactose agar plate and to a plate of 1 per cent dextrose agar. The broth tube showed no evidence of growth at the end of ninety-six hours. The tube was then placed at 10 C. for two weeks and then returned to room temperature. Within a few days a faint opalescence appeared. A few drops of the culture were then spread on a plain agar plate, and G colonies developed after ninety-six hours. These were typical G forms similar to those of the original culture.

The litmus lactose agar plate showed no sign of growth at the end of ninety-six hours or at the end of one month. The dextrose agar plate, however, although revealing no G colonies at the end of ninety-six hours at 37 C., showed a heavy sprinkling of very minute colonies (from 0.02 to 0.05 mm. in diameter) at the end of thirty days, and probably earlier. These colonies conformed in all respects to the G type culture that had been subjected to filtration nearly two years before. If any change had occurred in the culture, the recovered G colonies were smaller than those of the G culture that received filtration.

A considerable number of tubes of sealed filtrates were tested culturally during the few weeks beginning July 3, 1930. The method was the same as in the case of the sealed G cultures, namely, seeding to infusion broth, to litmus lactose agar plates and to plain infusion agar plates, and staining on slides by several methods, including that of Giemsa. The detailed results relating to the first twelve filtrates examined are embodied in table 19.

From table 19 it appears that none of the broth tubes inoculated with sediments of sealed filtrates showed clouding at the end of twelve days. Three agar plates similarly inoculated, however, revealed G

TABLE 19.—*Results of the Examination of Twelve Berkefeld Filtrates of G Type Cultures Sealed in Ampules for About Two Years*

Tube*	Date of Sealing	Appearance of Tube	Results of Cultivation†	
			Agar Plate	Broth
I-3A-c	3/12/28	Cloudy +; granular sediment	+	—
I-5A-c	3/12/28	Hazy; granular sediment	—	—
I-7A-d	3/14/28	Clear; granular sediment	—	—
I-9A-f	3/21/28	Clear; granular sediment	—	—
IV-II-3a	5/ 2/28	Hazy; granular sediment	+	—
IV-IV-3a	5/ 2/28	Clear; granular sediment	+	—
I-7A-b	3/14/28	Hazy; granular sediment	—	—
I-9A-g	3/21/28	Clear; granular sediment	—	—
I-21A-c	5/16/28	Clear; granular sediment	— +#	—
I-23A-b	5/21/28	Clear; granular sediment	—	—
IV-II-5a	5/ 5/28	Clear; viscous granular sediment	—	—
IV-IV-5a	5/ 5/28	Clear; granular sediment	—	—

* The first six ampules were opened July 3, 1930; the last six, July 9, 1930.
† Cultivations were in broth and on agar plates.
# Positive on plate made July 16, 1930.

colonies on the first plate seeded. Filtrate I-21A-c gave G colonies on the second test. Although the number of recoveries of the G forms in this series was small (25 per cent), we are inclined to believe that had the serial plate method been instituted in the cases in which results were negative, a larger number would have given positive results. The broth tubes are being kept for continued observation.

CHARACTERISTICS OF G ELEMENTS IN CULTURES RECOVERED FROM SEALED TUBES

Morphology.—The G colonies appearing on the plates seeded with the sediments of old sealed G cultures (or of G filtrates) were transferred to plain infusion broth and grown for twenty-four hours at 37 C., then examined in stained preparations (fuchsin, Giemsa stain). The following record for the ampule of G strain I (I-28A-a) is sufficient to give a picture of the morphologic types present in all of these cultures, since there was no appreciable variation at this stage of development.

The organisms were entirely coccus forms of some grade; no typical rods were present. The size ranged from 2 microns to the limit of vision. The largest elements were usually single. Those of intermediate size were more commonly in diplococcus or short chain groupings. The smallest were organized in long, thin, streptococcus chains. In these chains, the size of the individual elements was very small, and for the most part uniform, but occasionally larger cocci stood out from those adjacent. The coccal and granular elements in these chains were often widely spaced, the intermediate portions being made up of a pinkish staining sheathlike structure which was continuous. All of the forms here described stained fairly heavily, and we believe that they were living.

On the other hand, in the background of these slides, the field was covered with long streptococcus filaments, light pink, in which were embedded faintly blue-staining granules. We suspect that these forms were nonliving. If, however, these background chains and filaments are included, it seems safe to say that the predominant morphologic type of these cultures at a slightly earlier period was a streptococcus. In further development it transformed into the micrococcus and diplococcus stage. In some of the young G colonies, however, a few streptococcus chains persisted, as was shown by cover glass impression films made directly from the agar plates. It seems probable that the viscous nature of the sediments in the old sealed tubes of culture or of filtrate was due to this streptococcus grouping. This was also indicated from stained preparations made from the sediments themselves. In a few cases, it should be added, coccobacilli, and even short rod forms in small number, were observed. These forms increased with further propagation of the G cultures.

One interesting feature of the morphology of this and other cultures arising from the sealed ampules was that the slide pictures were almost identical with those of the broth cultures of the same G strain at the time of their first isolation more than two and one-half years earlier.

Filtration Tests on Recovered G Cultures.—In earlier pages it has been shown that the filtrability of the G type cultures was not an occasional feature, but a constant and invariable characteristic. It was a

matter of interest to ascertain to what extent the G cultures that had
been sealed in ampules for two years or more retained this characteristic. The morphology of the elements in stained preparations of these
"recovered" cultures suggested that they should be filtrable, or at least
should contain the filtrable elements. Of the G strains recovered during
July, 1930, two were subjected to the filtration test; both gave positive
results. The history of G strain I-28A-a will be presented first.

The original strain was produced in November, 1927, by enforcing the dissociative reaction on an R type Shiga culture (previously derived from an S culture)
by means of lithium chloride. During the spring of 1928 this strain received
many filtrations in order to test the constant nature of this characteristic. On
June 14, 1928, sealed ampules of the N filtrate were prepared and stored in the
dark at room temperature until July 3, 1930, when one ampule was opened and
examined culturally. Plating revealed the G colonies, from which a broth culture
was set up. This was filtered on July 14 through a Berkefeld N candle under
200 mm. of negative pressure.

On recovery, the filtrate was divided among several sterile tubes and incubated.
One portion was examined directly after filtration by spreading 10 drops (0.3 cc.)
on a litmus lactose agar plate and on a plain agar plate. Both were incubated at
37 C. for about two weeks. The litmus lactose plate was negative for eight days.
On the ninth day, examination with a $10 \times$ lens showed minute points that
suggested G colonies scattered thickly over the surface. Further examination
with the binocular, under reflected light, confirmed the observation; many thousands of minute G colonies were present. Over the next few days these colonies
increased slightly in size, so that they could be seen with a $3 \times$ magnifier. From
these colonies the G culture was recovered.

In the second case, use was made of G strain I-34A-b. This was recovered
from an ampule of G culture sealed on Aug. 1, 1928, and opened for examination
on July 3, 1930. The recovered G culture was grown in broth and again filtered
through a Berkefeld N candle on July 14, 1930. This filtrate, when seven days
old, was seeded on litmus lactose agar plates and transferred twice serially at
intervals of two and five days. The G colonies again appeared on the third serial
plate after forty-eight hours' incubation at 37 C.

From this record it is clear that the characteristic of filtrability was
maintained by the first G culture mentioned for a period of more than
two years in the sealed ampule. In this instance, it required nine days
for the colonies arising from the filtrate to become visible under the
$10 \times$ hand lens, and they never became clearly visible to the naked eye
until propagated further. In the second case, it was sixteen days after
seeding the filtrate that the G colonies first became apparent. We may
therefore conclude that, under the conditions of these somewhat pro-
longed experiments, the filtrable virus phase of the Shiga bacillus main-

tained itself as a highly stable body, in fact rivaling the stability usually accorded by bacteriologists to the ordinary forms of the Shiga culture.

Influence of the Bacteriophage on Recovered G Cultures.—On an earier page it has been noted that the original S and R type cultures from which the G forms were first derived were susceptible to the influence of the Shiga bacteriophage. This bacteriophage had no influence, however, on the derived G cultures; they were not modified by it in any way. On the other hand, when these G strains had again been transformed to the original types, they were again susceptible to the lytic influence. In view of these facts we were led to test the resistance of the G strains recovered after their two years or more in sealed ampules. Plates were spread thickly with the G cultures, and then a loop of one or another of recently prepared lytic filtrates active on *B. coli, B. typhosus* and *B. dysenteriae* Shiga was deposited on each of the plates. After incubation for eighteen hours, the cultures showed normal, uninterrupted development. Control plates seeded with the cultures homologous to the bacteriophage strains employed revealed a broad area of lysis. From these results we conclude that the long period of sealing had produced no change in the reaction of the G strains to bacteriophagic influence.

SUMMARY OF SECTION AND CONCLUSIONS

In the present section we have reviewed what is perhaps one of the most interesting and significant characteristics of the G forms and their closely related filtrable elements—namely, their viability and longevity under unfavorable conditions. Cultures of the Shiga dysentery bacillus are notoriously difficult to keep alive in the culture tube even over the period of a few months. But we have shown the existence of a special form of this organism which, imprisoned in glass-sealed tubes, without adequate food materials, surrounded by its own metabolic products and presumably in the presence of a minimal supply of oxygen, easily maintains its existence for more than two years and probably still longer.

The data that have been presented on the sealed cultures of the G forms indicate that, over this period of time, they undergo little if any change. Indeed, it is suggested that during the long period in the sealed state they manifest a greater tendency to revert to the noncultivable virus form than to undergo progressive development into the normal S type Shiga culture. Moreover, it is indicated that the long survival of the culture is by no means dependent on the continued existence of a bare remnant of the G elements, but on the contrary is characterized by the presence, in every drop of culture sediment, of thousands of minute living bodies.

The data presented on the sealed filtrates are sufficient to demonstrate that, in a similar manner, the filtrable virus form of the Shiga organism not only maintains itself as such over similarly long periods (as shown by the filtration tests performed on 2 year old sealed filtrates), but, in varying degrees, may accomplish a progressive evolution into the visible and cultivable G forms (as indicated by the results of direct platings from the sediments present in old, sealed ampules of filtrate). It is perhaps no less remarkable that some of the virus forms survive as such than that others retransform into the cultivable G elements and again become capable of registering on agar plates in the form of the minute G colonies.

All of these facts, whether concerning old, sealed G cultures or old, sealed G filtrates, emphasize the point that, in nonspore-forming bacteria, it may be the minute G elements, or perhaps the still more minute virus forms, that serve to maintain the life of the species under conditions that are unfavorable for continued, vegetative cell multiplication. This view becomes the more understandable, perhaps, if we regard these minute elements as the true fructification bodies or "seed forms" of bacteria.

COMMENT [19] ON RESULTS

In the introduction to this paper it was stated that the demonstration of the existence of special culture types entitled to the term "filtrable forms" of the organism might be significant from several points of view. We believe that the evidence submitted demonstrates the existence of such a filtrable stage in the life history of the Shiga dysentery bacillus. And it may be repeated that one of us (P. H.), in collaboration with other workers in the laboratory, and others, working independently, have already produced analogous filtrable forms in ten or more other bacterial species. Some of the studies mentioned may be reported at a later date; but it may be said that, so far as the investigations have progressed, they confirm in all important respects the observations now reported for the Shiga bacillus. We are therefore in little doubt that the phenomenon is one of general significance; in fact as much so as that of the S and R types, which are now rapidly

---

19. The senior author alone is responsible for the conclusions advanced in this section.

becoming recognized for a considerable number of species. For this reason we feel justified in examining in a tentative way the possible significance of these results for certain problems related to bacteriology, pathology and medicine.

NATURE AND BIOLOGIC SIGNIFICANCE OF G TYPE CULTURES

First, however, it seems desirable to summarize briefly some of the major points mentioned earlier, and to consider with what justification the G forms may be regarded as the filtrable and cultivable stage of the Shiga species. We renew this aspect of the problem at the risk of being tiresome to some of our readers. But we feel that the notion of "contaminations" in connection with filtration studies has been so forced on the attention of bacteriologists that a certain timidity has developed both with reference to attacking filtration problems and with relation to interpreting the results when obtained. It is for this reason particularly that we are so insistent on dealing fairly with this problem of "contaminations."

The first possibility is that the G cultures might be contaminations which gained access to our stock cultures after the series of platings made for the isolation of pure line S and R stocks were performed. In this case, they are contaminations that appear to be practically identical in the fifteen or more instances in which they have arisen in our Shiga stocks. They are contaminations which have entered pure line S and R strains that, in the course of ordinary cultivation in broth or on agar, reveal no sign of their presence. They are contaminations which are brought out particularly by the use of lithium chloride, pancreatin or the bacteriophage. They are contaminations to which both young and old cultures are susceptible; and which, with their appearances and disappearances, play hide-and-seek with the investigator. If they represent contaminations in our tubed mediums, they are forms which in some cases withstand heating at 15 pounds (6.8 Kg.) pressure for thirty minutes. In the serial cultivations of S and R cultures in lithium chloride and pancreatin broth, they are contaminations which fail to reveal themselves by common methods of plating from the earlier tubes of transfer; they do not make their appearance until a certain "moment" in the series has arrived (usually the sixth to twelfth tube), and then they appear suddenly and in great numbers. If they are contaminations in our filtrates, they are contaminations which, when present in fresh filtrates, may reveal, over a long series of plate to plate transfers, none of the common signs of culture growth. And, finally, they are contaminations which, by some curious circumstance, may become slowly transformed into cultures indistinguishable from those in which they first arose.

But there are probably some conservative bacteriologists, particularly those clinging to what little remains of the monomorphic conception, who will suspect that the filtrable form to which we have thus far referred as the G type culture is in reality a distinct biologic entity—not necessarily a laboratory contamination in the ordinary sense, but a strange microbe living in close association with the Shiga (and perhaps other) cultures. This notion would be in harmony with a view that has gained some popularity in recent times—as, for example, the notion of the bacteriophage in the guise of a foreign, filtrable virus, living "in symbiosis" with either filtrable or nonfiltrable and resistant stages of the culture of the substratum (d'Herelle and Hauduroy). Or, again, the "filtrable virus" of scarlet fever, living "in association" with the alleged scarlet fever streptococcus (Zlatogoroff and others [1929]). This is a form of objection that is more difficult to meet, and, theoretically at least, the view is probably proof against controversion. The only effective method of dealing with an argument of this sort would be to demonstrate that, wherever the "virus" of scarlet fever exists, there also stands the filtrable stage of the streptococcus; and, conversely, to demonstrate that, wherever the filtrable form is absent, there is no "virus" present. Under these conditions the alleged association becomes (for practical purposes, at least) a potential identity.

If this is not the case, then we must assume that the associate lives in such close relations with the principal culture that repeated colony selections fail to purify the strain. We may even go so far as to say that the contaminating species is able to form minute, independent, secondary colonies within the larger colonies of the original culture. We have already seen how, in the cultures of the typhoid bacillus studied by Miss Bailey, the G type colonies appeared as secondaries inside the old, "normal" S colonies shortly before they registered in numbers as discrete G colonies on the plates. This implies to us that the S or the R type cells, at a certain moment in their development and at that time only, are able to generate, or to become transformed into, the G bodies or their prototypes. This phenomenon does not appear in every culture and it may be necessary to wait for many tube-generations to pass before the critical moment of transformation arrives.

It is probably unfortunate for our thesis that the G forms of all the cultures studied in this laboratory, with the exception of those of B. anthracis, obtained and studied by Mrs. Dorothy Parker, resemble each other so closely in point of morphology and colony structure. This circumstance seems to afford evidence for the hypothesis of "universal contaminant." But we may bear in mind that many species of micro-cocci are much alike morphologically; that the "seed forms" of fungi often resemble each other much more closely than the adult forms. The full expression of the individual or the species characters is attained only through the course of ontogenic development, to which we believe the cyclogenic progress of the bacterial forms shows somewhat of a parallel. Many diverse forms of living things meet on a common foot-ing in the embryo and even more so in the "seed." These comparisons, in the present case, will seem to many as highly allegorical; but it should be recalled that there are certain biologic laws of function and develop-ment that are never overstepped, whether one is dealing with microbes or with men.

Against the hypothesis of a "universal contaminant," however, some-thing more than theoretical biologic deductions exists in our favor. If the G type cultures of the Shiga bacillus, the typhoid bacillus, the para-typhoid bacillus, B. enteritidis and all the others in this group possess such marked similarity, what might be expected of the presumably existent G forms of a species so remote as the bacillus of anthrax?

Among the various types of anthrax cultures observed by Nungester in his recent investigation in this laboratory, there was one that he termed the "midget" colony type. It arose commonly from his "smooth phantom type." The "midget" forms were not studied in detail by Nungester, but have been investigated in a preliminary way by Mrs. Parker (personal communication to the authors). These forms have now been found to be filtrable; at least, they give rise to minute bodies that easily pass the Berkefeld N and W candles. On seeding the fresh filtrates to agar plates, no growth has commonly appeared earlier than the third plate in the series. There then arise minute colonies that seem to be analogous to those seen in the Shiga cultures. But this difference is to be observed: The minute colonies of B. anthracis reveal a distinct and characteristic morphology that is in keeping with the morphology of the anthrax organism. This observation is perhaps suffi-cient to afford the basis of a suggestion that, although the G colonies and cultures of closely related species appear practically identical, the analogous forms of distantly related species will be found to possess their own characteristic features which are readily distinguishable, one from the other.

For this reason and for others that have been brought out in earlier sections of this paper, we believe that there is no justification for assuming a new species to meet the requirements of the present case. The rapid increase in the registrations of new bacterial "species" and of "mutants" each year in the archives of the science is largely due, we believe, to a failure on the part of bacteriologists to study and to recognize the complete biology of the established species for which more than adequate names are already available. And it can scarcely be doubted that when the complete biology of the recognized species is known the number of "good species" catalogued will be reduced in considerable degree.

If, as we believe we have demonstrated, the G forms do not represent a contamination or a foreign species closely associated with the Shiga cultures, but are produced by it and within it as one of its constituent cyclostages, one of the first questions to arise is: Are these forms accidental pathologic variations ("impressed variations") that arise under the harmful influence of lithium chloride or of pancreatin, or are they "normal" forms that play an important rôle in the life history of the species? The former view is the one that will probably appeal most strongly to the majority of bacteriologists. That is, the G forms (colonies or cells) are stunted variants that demand merely a return to favorable environmental conditions to revert to the original or "normal" type. This view is in line with the older conceptions of monomorphism and its associated "involution forms," "degenerate forms" and "pathologic cells" in general. Some evidence, however, can be brought against this conception.

First, it may be noted that one of the early discoveries of the G type culture was made in connection with an agar plate that had been seeded from the tenth tube in a series of plain infusion broth cultures. Moreover, it has been established in other cases in which the G type cultures of other bacterial species have been produced in our laboratory, that they often arise in young cultures growing in a plain broth medium. Although the number of instances in which success was attained from the use of lithium chloride was relatively greater, it nevertheless must be concluded that this substance, as also pancreatin, is able to accomplish, in culture modification, no results that cannot be readily obtained in plain broth mediums and with young cultures.

Another reason why it seems unwarranted to regard the G forms as degenerate or dwarfed cells is that continued propagation in the same medium containing the "injurious" substances may bring about a retransformation to, or at least a return of, the original culture type. Here is involved, however, a point of considerable significance related not only to the generation and subsequent disappearance of the G forms, but also to the S to R and R to S transformations. It is therefore necessary to deal briefly with this subject.

It has become somewhat common, of late, to regard microbic dissociation as comprising a transition from the more commonly known culture form of any species (usually the S) to the SR, R or O type; and this aspect of the dissociative reaction was probably given too great emphasis in an earlier work on the subject by one of us (P. H., 1927)—the reason being that the S to R transformation was the one more commonly recognized at that time.[20] But, as we have

20. In this connection it is of interest to note that the terms "dissociated" and "dissociate" possessed a somewhat different and more limited meaning when they were used in the earlier days of bacteriology. The original meaning of the term "dissociated form" was the reverse of the usual current meaning. Originally, it referred to the breaking up of cultures, already in a filamentous stage, into short bacillary or even coccoid forms, and was never employed to denote the reverse transformation, which was known equally well to occur. The term was applied particularly to those organisms, for example, Streptothrix, Leptothrix and Cladothrix, the thread forms of which were better known to bacteriologists than the short bacillary forms, which were manifestly produced from the threads. The change therefore represented a "dissociation" from the better known, "normal" forms to forms less well known. At present, the same may be said to hold true, although the sense is now reversed, since in these days the separate bacillary bodies of most bacterial species are better known than the related thread (fungoid) forms of the same species. Since the term "dissociation" has the connotation of breaking up into smaller particles or aggregates, the earlier usage possessed the more logical application.

Among the early works in which the term "dissociated forms" is used may be cited that of Vicentini (Bacteria from the Sputa and the Cryptogamic Flora of the Mouth, London, 1897) ; also the interesting study of Billet (Contribution a l'étude de la morphologie et du développment des Bactériacées, Paris, 1890). The same sort of phenomena were mentioned and described still earlier by Ray Lankester (1873), Zopf (1881) and Cienowski (1877). Indeed, the conception of bacteria as representing a botanical group appears to have been much more clearly defined and utilized in the preceding century than it is today.

shown by many experiments involving the use of lithium chloride, pancreatin and other reagents, it usually happens that the substance (lithium chloride, for example, which has been more fully studied by us in this respect) readily forces the change either from S to R or from R to S. But the trend of culture modification, in either case, does not always terminate at the end-point mentioned; for, after the S to R transformation has been effected in such a medium, further serial passage of the same culture in the same medium determines just as readily a reversal from R to S. Thus, continued passage from tube to tube may, if the reaction of the medium is favorable (and this is very important), determine several transformations and reversals, some of which may be complete (as observed from the differential colony count) and some incomplete. The same sort of transformation and reversal may be observed, over long periods, in relation to the S and the G forms. If the G form is not isolated when it first appears, continued mass cultivation in series of the culture in which it first arose often leads to an apparent disappearance of the G elements after a few more transfers. They may reappear, however, in some later tube in the series, then disappear and appear again. Occasionally the G type culture alone survives, and the culture mass seems to undergo no further change. Such a picture of transformation and retransformation, as manifested by the S, R and G forms, can scarcely be interpreted as representing the production of degenerate cells under the malign influence of foreign substances in the medium; for, in such event, it would be necessary to suppose that the stunted or otherwise degenerate cells were brought back to normality by a continuation of exactly the same influence that first accomplished their degradation.

From the results in hand it seems more logical to conclude that "dissociation" involves merely transition, and that this may occur in several directions; that moreover, the influence of lithium chloride (and presumably other substances, including homologous immune serum, some normal serums and peritoneal exudates) is not such as to inflict a permanent disorganization on the cells, with the result of producing and maintaining pathologic forms, but such as to set into operation or to intensify, in the culture mass, its normal and innate tendency to proceed along the path of its specific cyclogeny; and that in this process it gives rise to many cell forms, some of which, in small numbers, are perhaps always hidden in the culture, but which, when brought to the front and segregated in the pure state, appear to us as strange and incomprehensible.

For these reasons it seems hardly logical to attribute the generation of either the G or R type exclusively to the "injurious" effect of chemical substances in the mediums. We must therefore turn to other possibilities in order to discover a satisfactory explanation of the significance of the G forms. This will necessarily lead us into the newly developing field of bacterial cyclogeny, on the clear recognition of which not only the significance of the G type cultures, but also the explanation of many other curious aspects of bacterial behavior, must eventually be based. This phase of the problem will be considered presently. First, however, we must deal briefly with another matter of importance.

RELATION OF FILTRABLE FORMS OF SHIGA BACILLUS TO VISIBLE ELEMENTS OF G TYPE CULTURES

Is the filtrable virus stage of the Shiga bacillus identical with the visible elements present in the G type culture, or are the filtrable bodies merely associated with this form of culture? This question cannot be definitely answered at present, but some facts bearing on it may be considered.

First may be mentioned the relation between the grade of candle employed and the time when the G colonies first appear on the plates seeded with the filtrates. We have already noted that the bodies that pass the N and W candles are of such a nature that their regeneration into the visible G culture does not occur, as a rule, until the fresh filtrate has been passed over from three to six agar plates, while filtrates that are derived from V candles generate the G forms somewhat earlier in most cases and frequently on the first plate. In all cases, however, irrespective of the grade of candle employed, the G culture is eventually obtained from the filtrates. Moreover, this is the only form of culture invariably recovered from filtrates derived from intact candles.

These results might suggest that the form of the organism passing the N and W candles is not commonly one that is capable of immediate cultivation on agar plates; or, if it is so cultivable, it is not in a form that is readily detected or that subscribes to our common notions of culture growth. Moreover, these results suggest that the form of organism (or one of the forms) passing the V candles is directly cultivable, often yielding typical G colonies on the first plate seeded. These observations, in turn, might imply that the elements passing the N and W candles are different from some of those that pass the V candles; in other words, that the elements passing the finer grades of candles are analogous to those passing the V grade only after the former have undergone some sort of development toward the microscopically visible G type culture. Thus it might happen that the G culture, which we can plainly see, is not made up altogether of filtrable bodies, but that a certain number of these bodies are associated with it.

It is the same apparent evolution of the filtrable elements into a visible culture form that undoubtedly underlies another phenomenon in Shiga and other species, next to be mentioned.

If the filtrable elements present in fresh filtrates are incapable of producing visible growth on the plates until several serial transfers have occurred, or until several days have passed, it would seem probable that "aging" the filtrates would determine a similar development. This is what actually happened, as has been demonstrated by several tests. The results have shown that filtrates aged for from six to ten or more days are more likely to yield G colonies on the first plate seeded than are the younger filtrates. Here, then, is additional evidence that a development of the filtrable bodies occurs in stored filtrates, just as it does in the course of passages across agar plates. The filtrable bodies gain the characteristic of cultivability in visible form, and after about the same length of time, whether the "aging" is carried out on agar plates or in the original filtrate tube.

The considerations just mentioned might suggest the existence of two forms of filtrable bodies: (1) those that usually pass the V candles and readily manifest themselves on plates directly by revealing the G type colonies; (2) those that pass the N or W candles (undoubtedly the V also) and manifest themselves on plates only after several serial transfers. It would seem to us more reasonable, however, to picture the filtrable bodies as representing, not two fixed types, but a series of continuous variations extending from the W-filtrable to the V-filtrable elements. Since, as we could easily observe, there exists a very slow and gradual merging of the G forms into the "normal" form, there is likely to exist a similar evolution from the filtrable elements into the microscopically visible and cultivable, but distinctly rudimentary, G type culture. At present, however, this is not capable of experimental demonstration, other than that already reported. We incline to the opinion, however, that our results give some support to the view that the bodies that are the most readily filtrable through the finer grades of candle represent a "resting stage" in the development of these elements, and that it is only when this stage has passed that these virus-like bodies possess the power of yielding visible growth in or on any of the culture mediums.[21] When growth first becomes possible, it occurs only in liquid mediums (or on the surface of moist plates); later it may occur on ordinary, dry agar, then giving the picture of the G type culture. When, on the other hand, the filtrates are able to yield at once a visible growth on the plates, it seems probable that this is because the coarser grade of candle permits to pass, not only the resting stages (virus-bodies), but also some elements(perhaps larger) that have passed by this stage and consequently are able to yield at once a visible growth on the plates. That there exists at least some sort of "latent stage" of development of the elements present in the majority of N and W filtrates cannot be doubted. These observations, as well as our conclusions, are in harmony with certain conceptions presented by Löhnis (1923) for *Azotobacter* and by Enderlein for several bacterial species. They are also in harmony with all that we know regarding the regeneration of gonidial (conidial) bodies in the higher fungi.

Briefly to summarize our conclusions dealing with the subject of the present section—we believe that the data presented support the following view: The G type culture does not comprise the filtrable virus form of the Shiga bacillus, but includes it, at least to a much greater degree than any other culture type of which we possess any knowledge. The filtrable elements, at a certain stage of their development, do not manifest visible growth in or on any culture medium. In due time, however, and very gradually, they generate the visible granular and coccoid elements of the G type culture, which itself may be a growth of extreme delicacy. This type of culture, for a considerable time at least, may not only maintain itself, but also regenerate more of the filtrable bodies. Beyond a certain point in the development of the G culture, however, this ceases to be the case, and the culture begins its gradual ascent, through ontogeny, toward the "normal" type, and subsequently to the reproductively mature form. According to our view, the G type culture may thus be regarded as the intermediate culture state lying between the filtrable virus stage of the organism and the adult form.

---

21. On the other hand, we acknowledge the difficulty of differentiating between such "resting stages" and a virus form that is multiplying, but without visible evidence of it.

POSSIBLE RELATION OF G TYPE CULTURES TO OTHER CULTURE
FORMS REPORTED IN THE LITERATURE

The question now arises: Does the G type culture represent a form that is new in bacteriology? We much doubt that this is the case; for we believe that it has appeared countless times in the experience of bacteriologists, but that its significance and that of the phenomena accompanying it have, in most cases, remained unappreciated. The minute colonies, when observed, have usually been regarded as colonies that manifested some sort of arrested development. They were not regarded as a colony type essentially different from the normal type. Until recent time, it may be recalled, there has been but slight general recognition of the S and R colony forms; and even today these are a source of anxiety to many workers, especially those engaged in the field of systematic bacteriology.

But the notion of the existence of a minute stage of culture development, resembling the gonidial bodies of the fungi, has been dawning in the bacteriologic world for many years, and we have pictured in our G type cultures and their associated filtrable forms a morphologic type that we believe to be equivalent to the gonidia of higher forms. In view of this interpretation, it now seems desirable to review briefly some of the more important of the earlier observations which have supported the view that bacteria are able to generate minute "seed forms" possessing the physiologic equivalence of the gonidia of the fungi.[22]

Although Ehrenberg (1838) stated that some of his "infusoria" produced "eggs" which were eventually liberated from the cells, it was not until the work of Perty (1852) that bacterial gonidia were clearly pictured. These he observed in certain species of *Spirillum*. He depicted the origin of these minute bodies inside of the spirilla and their subsequent development, after liberation, into the original cell type.

In 1870, Cohn pictured still more clearly the gonidial bodies in *Crenothrix polyspora* and gave a clear description. He mentioned the minute "microgonidia" and the larger "macrogonidia." Cohn did not, however, extend these observations to other bacterial species. It is

22. Many of the following references are taken from Löhnis' monumental work, "Studies on the Life Cycles of Bacteria." Here is to be found an extensive presentation of the literature dealing with modes of bacterial reproduction. The interested reader should consult it.

of much interest, as pointed out by Löhnis (1921) that Koch (1877), in some of the earliest photomicrographs ever taken of bacteria, showed (plate XV, fig. 4) what he termed "Bacillen mit mehren seitlichen Sporen." The organism here represented was said to be *Bact. termo*. The "seitlichen Sporen" manifested a striking resemblance to gonidial bodies budding from the sides of the rods. Again, in *B. anthracis*, Koch pictured (plate XVI, fig. 5) similar round lateral or terminal buds, but made no comment in the text regarding their nature or significance. If he had done so, bacteriology would have been different in many important respects. Similar coccus bodies in cultures of *B. anthracis* were pictured by Gunther (1906), but not mentioned in the text. We ourselves have encountered them in the process of formation in the majority of anthrax cultures examined, at some time in their development. They often gave the impression that the anthrax culture had been contaminated with a micrococcus; however, the origin and the destiny of these "micrococci" could be easily followed.

In 1874, Burdon-Sanderson stated that there occurred in some of his cultures certain "spheroids" which apparently developed into rod forms. He presented the possibility that bacteria might be formed from more minute and "ultramicroscopic" bodies.

Billroth, in 1874, stated that minute coccus bodies might play an important part in bacterial multiplication. His drawings (original figs. 36 and 42; reproduced in Löhnis' "Studies on the Life Cycles of the Bacteria," plate L, figs. 44 and 45) reveal the generation, multiplication and final liberation of the gonidial bodies; also sporangia-like bodies resembling the sporangia in Cohn's *Crenothrix*.

At this point, we must pass over the important observations of Cienowski (1877) on the production of gonidia by *Crenothrix* and *Cladothrix;* of Geddes and Ewart (1878) and of Ewart (1878, b) on *B. anthracis,* who observed the "spores" escaping from the sides and ends of the rods; of Israël (1878) on *Actinomyces;* of Albrecht (1881) on *Spirochaeta obermeieri,* and of Zopf (1879-1885) on the gonidial bodies of the *Trichobacteria*.

The work of Fokker (1881-1882) is of special interest, since he combated the views of Koch (1887) as to the specific relation of the anthrax bacillus to the disease. Fokker pointed out that often no bacilli, but only minute coccoid bodies, could be found in the tissues. These have often been seen in later years and as often discarded (when found in anthrax cultures) as contaminating elements. They are probably identical with the lateral and terminal buds observed by Koch and others.

According to Magnin and Sternberg (1884, p. 150), Toussaint once reported that the anthrax bacillus produced both large and small "sporangia" out of which from three to six "spores" escaped. He believed that he could observe the transformation from these "spores" to the bacillary form in cultures kept on the Ranvier slide (chamber). It is of further interest to note that, although Koch and others of his school were opposed to these views, they were ready to regard as "spores" the minute bodies observed in the tubercle and typhoid bacillus (Koch, 1882). Both Klebs (1883) and Babes (1883) were able to demonstrate, to their own satisfaction, that the granules were produced by the bacilli themselves, and that the granules, in turn, reproduced the bacillary form.

Here again we must pass over the suggestive observations of Neisser (1888) on B. xerosis; of Beijerinck (1888) on the "swarming bodies" of B. radicicola; of Dowdeswell (1889-1890) on the granules of the cholera vibrio ("sporangia" producing "sporules"), and of Mac Fadyean (1889) on the "seed forms" of Actinomyces. In the light of present knowledge, all of these observations manifestly dealt with the production of bacterial gonidia.

Lehmann (1888) observed that heat-resistant "microspores" were present in B. anthracis in his cultures, and this was confirmed by Roux in 1890. Novy (1894) noted that an asporogenous strain of B. oedematis-maligni resisted heating at 58 C. for one hour, suggesting that the granules present inside and outside the cells might be more resistant than the vegetative rods.

Zettnow (1891) presented photographs showing the budding out of regenerative bodies in bacteria and spirilla, but he paid no attention to these phenomena, referring to them merely as "globules" that separated themselves from the parent cell and became free, or as "krankhafte Degenerationsformen."

Cunningham (1897) showed that the cholera vibrio was reproduced from coccoid forms. By other workers the same was found true of the Finkler-Prior vibrio and V. metchnikovii. Migula (1897, vol. 1, p. 157), however, regarded all these manifestations as examples of contaminations, although it is of interest to note, as Löhnis (1921) pointed out, one of these contaminants appears in one of his own photographs, attached to the "back" of one of the vibrios. By Migula, reproduction of bacteria by the formation of gonidia was regarded as impossible.

Almquist (1893 and 1911) showed that both *B. typhosus* and *B. coli,* when kept at low temperatures (from 10 to 11 C.), produced minute round bodies which later reproduced the original form. In 1904, he observed larger round bodies which budded from the sides and ends of the vegetative rods, and which he termed "conidia." These conidia were able to multiply as such by budding. These observations were confirmed and extended in the years from 1906 to 1908. In 1916, he again pictured the "conidia" of *B. typhosus* and *B. dysenteriae.* The latter appear identical with some forms that we have described in the present paper.

In the spirochetes, many observations of "granular" or "globoid" bodies have been made subsequent to 1906, when Novy and Knapp first demonstrated the filtrability of these forms. From the works of Dutton and Todd (1905 and 1907), Leuriaux and Geets (1906), Krienitz (1906), Breinl and Kinghorn (1906), Wolbach (1915), Perrin (1906), Breinl (1907) Dutton and Todd (1907), Leishmann (1909), Balfour (1911), Hindle (1911), Fantham (1911), Nicolle and Blanc (1914) and others it becomes increasingly clear that the "granules" within the spirochetes may multiply as such, or may eventually become transformed into the typical spirochete. Such a view accounts for much that is now obscure in the behavior of the spirochetes in the living host.

In 1910, Fontès accomplished the filtration of the tubercle bacillus. This was confirmed by Vaudremer in 1923. The filtrable ("micrococcus") form was apparently cultivated as an independent organism by Panek and Zakharoff in 1930.

In 1913 and 1920, Jones pictured what he believed to be flagellated gonidia in *Azotobacter.* In 1914, Meirowsky pictured the gonidia of *B. paratyphosus* and *B. enteritidis.* Hort (1916, a) observed the production of gonidia by the large parent cells of the meningococcus. At the time of his observations, Hort believed this phenomenon was rare among bacteria, and for this reason he placed the meningococcus among *Hemiascomycetes.*

In 1916, Löhnis and Smith depicted the liberation of gonidia by many bacterial species, and in 1923 Löhnis showed them in *Azotobacter.* In 1922 appeared the monumental work of Löhnis in which he discussed in detail the methods of reproduction of bacteria, presented numerous photomicrographs to substantiate his claims in respect to a considerable number of bacterial species, and presented a comprehensive review of the literature dealing with gonidial bodies and other curious bacterial structures to the time of his publication. A thorough reading of this work is essential to any one who would have complete knowledge of the status of the problem of the nature of bacterial reproduction.

In 1925, Enderlein, in his comprehensive work on bacterial cyclogeny, again brought into the foreground the bacterial gonidia with special reference to their origin and multiplication. Enderlein's study dealt more with the cytologic than with the cultural details, and introduced new subjects of much significance relating to the bacterial nucleus and its behavior in reproduction processes, which he fully described. The contributions of Löhnis and of Enderlein serve to establish the new and important subject of comparative bacterial cytology, the details of which, we believe, will henceforth enter into many aspects of bacteriologic thought and practice. This subject appears to represent the groundwork of many curious phenomena of which the manifestations of microbic dissociation are merely the superficial indications.

In 1927, Haag published his important study on the biology of the anthrax bacillus, presenting for the first time a fairly complete account of its cyclical development. In this paper, he described and pictured the gonidial forms which play so important a part in the regeneration of the anthrax culture. He indicated their modes of origin and their liberation from the parent cells. His observations go far in making clearer some of the curious facts noted in cultures of this organism, beginning with the period of Koch.

The work of Haag, from Lehmann's laboratory, was followed in 1929 by a similar biologic study of *B. mycoides* by Oesterle and Stahl. Among other interesting cyclostages, including the R and the S, these authors depicted the filtrable forms and the gonidia of this species; also large round or oval cells which they regarded as the "gonidangia," following the interpretation of Löhnis (1921). The filtrable forms were produced under the influence of sunlight, sodium chloride, soda, mercuric chloride and chloramine-Heyden; also, in some cases, by exposure to ultraviolet radiation. They appeared as minute granular or coccoid bodies which grew poorly on culture mediums. Eventually, however, they reverted to the original culture type, manifesting several intermediate growth forms.

The foregoing observations, covering the earlier period of bacteriology when, as we delight to say, methods were crude and exact observations perilous, might suggest that, in reality, bacteriologists of fifty years ago knew considerably more about bacterial gonidia than is the case today. It is perhaps possible that, in those days, when there was so little to do with bacteria after they had been isolated, eyesight was keener and microscopic examination of bacterial cultures more highly estimated. In any case, we regard these earlier observations, and many others that could be presented if space were available, as supporting our conclusions regarding the existence of a special form of reproductive body in bacteria, which it is permissible to regard as the gonidium. These bodies arise in various ways from the previously existing cell forms, and we believe that their significance and importance will become reestablished in the bacteriology of the future. If one is not pleased to regard these bodies as gonidia in the sense of "seed forms," as in the fungi, it does not greatly matter. Further study may reveal distinct differences. It is natural, however, in the examination of unfamiliar elements of this sort, to seek some analogy in the ontogeny of the most closely related biologic forms. In the present case it is especially appropriate to seek the analogy among the fungi, in view of the circumstance that much of the life of a bacterial culture in the "wild" state is spent in the fungoid stages of its existence. These stages embrace particularly the R forms of culture, which, as a rule are less familiar to bacteriologists than the S, since the latter seem, in the majority of species, to be best adapted to life in or on artificial culture mediums.

These fungoid stages are of much interest to those bacteriologists who have worked with them, although many seem still to doubt their existence. But the fact remains that there also exists, in the complete cyclogeny of many, and perhaps all, bacterial species, another stage that is just as definite and clearcut; a stage characterized by minute, and usually filtrable, coccus-like elements, to which some name must be applied.[23] It is to this class of bodies that, we are convinced, the filtrable elements described in this paper belong.

Among these elements, we have referred to some as "gonidia" and some as "microgonidia." Although we are not sure that the botanical literature indicates the existence of any qualitative difference between these two forms, it may be added that we have used these terms in the following sense: We regard the gonidia as the relatively large coccal elements that are first liberated by buddings or by "gemmulation" from the parent cells or filaments of the culture. These may regenerate, in

---

23. The terms "infrabacteria," "arthrobacteria," "protobacteria" and "antebacterial forms" used by certain authors are perhaps harmless enough in the present state of knowledge, but it must be admitted that they tend to obscure what we believe to be the true meaning and significance of the filtrable forms of bacteria.

the course of a few generations, the original cell and culture type. On the other hand, it is also apparent that these bodies, instead of at once instituting a reversion to the original form, may continue to divide into smaller and smaller elements before the reversion finally occurs. It is indicated that some of them lie below the range of ordinary microscopic vision. It is these smallest elements that we refer to as the micro-gonidia. We have seen evidence of these accessory divisions in the gonidia of the Shiga bacillus and the anthrax bacillus, and they have been reported often, in the older literature, for the cholera vibrio and some other species (Löhnis, 1921).

RELATION OF G TYPE CULTURES TO MICROBIC DISSOCIATION AND TO SPECIFIC BACTERIAL CYCLOGENY

From the facts already presented we have been approaching the conclusion that the G type cultures are a strictly normal product of dissociative activity in the mother culture; further, that they represent, in the bacterial development, a definite cyclostage which seems to show some analogies to the gonidial forms of the higher fungi. Although these forms apparently may sometimes be present in small numbers in "normal" cultures, it is only at certain stages of culture development that they become so numerous as to attract special attention. At such times, they may even represent all that seems to comprise the living culture, to the exclusion of all those organisms that have long been regarded as typifying the culture in question. This is particularly true of very old cultures. Cultures in this state no longer give the expected signs of colony or mass culture growth; sometimes they may be, to all appearances, sterile. But, in reality, they may be far from sterile. They may be swarming with living, and sometimes actively propagating, units, which are as numerous as the "normal" organisms in the culture from which they arose. Their subsequent development, when transferred to broth mediums and propagated in series, may for some time be such as to escape ordinary observation. When seeded to agar plates, they may reveal none of the appearances that we have been taught are the necessary accompaniments of bacterial growth; the plates are regarded as sterile. Some of these forms, or forms closely related to them, pass in considerable numbers the Berkefeld V, N and W candles; and they, or forms into which they develop, may invariably be recovered from the filtrates by cultivation methods, if the adequate technic is employed. If not, they remain hidden from view.

As pointed out in the section on origin of the G type cultures and in tables 1 to 5, the first appearance of the G colonies seems to be related to certain more or less orderly transitions between the S and the R cyclostages, frequently following or accompanying the disappearance of the other elements. In some cases, the G forms disappear and are followed by the S type cells. In some cases, the plates appear sterile for one or more transfers immediately preceding the first appearance of the G forms. How do these forms arise? Why, in continued mass transfer, do they sometimes appear to eliminate the other culture forms? What is the fate of the cells that serve for the generation of the G elements? Why do some G cultures remain stabilized over long periods, while others quickly revert to the original type? Why do some of the G cultures grow luxuriantly in broth, but refuse to give visible growth on agar plates? No entirely satisfactory answer can be given at present to these questions. All that can be said is that observations on the morphology of these forms, on their apparent origin in some cases, and especially on their subsequent development combine to suggest that they comprise a definite stage in the ontogeny of the Shiga species. We therefore accept the view that they are partially stabilized, and to an extent cultivable, gonidial forms (or elements representing their earliest stages of regeneration). They thus manifest a close analogy to the "Fruchtformen" of the higher fungi,[24] first recognized, so far as the relatives of the bacteria are concerned, and named, by Cohn (in *Crenothrix*) in 1870 and 1872, and later described by several other workers in the early days of bacteriology.

In this connection the question arises as to the justification of our elevation of the G form of the Shiga bacillus to the status of a new and special "culture type."[25] In answer it may be said that the present

24. Within the limits of this paper it is impossible to examine the details of comparative bacterial cytology on which this view is based. We can only cite the earlier observations of Cohn, Lister, Finkler, Prior, Almquist, Hort, Enderlein, Löhnis and Mellon, as furnishing the background for our interpretations of the many cytologic pictures that we are unable to present within the limits of this work.

25. It seems probable that the forms of culture occurring in the course of microbic dissociation and now termed "culture types," such as the S, R and O, will in the course of time receive different designations—names in harmony with the biologic significance of these forms. In the meantime, the designation of these clearly distinct stages of development by letters serves the purpose of making possible easy and concise reference.

basis for the recognition of the distinct culture types already well established, as the S, the R and the O (when found), involves mainly three points: (1) colonial form, (2) antigenic constitution (as manifested by serologic tests) and (3) degree of stability in pure culture. The evidence already presented in these pages demonstrates that the G type culture gives a colony picture that is distinct from all the others. We have seen, moreover, that this form enjoys a stability at least as great as that of the R form, and considerably greater than that of the S form. In order to complete the picture (although it has not been our intention to discuss the serologic aspects of the G type in this paper) it may be added that some preliminary tests have indicated that this form possesses antigenic and serologic characteristics that distinguish it from both the S and the R forms.

To some workers this result will appear in the light of an additional argument against the acceptance of the G type cultures as a stage in the development of the Shiga species. To such an objection, however, it must be reiterated, as one of us (P. H.) has shown in an earlier paper (1927), that there is no justification for the assumption that symmetry in serologic reactions is always a proof of the identity of species or, on the contrary, that a lack of such symmetry is always a proof of the difference of species. For the present it suffices to state that, when judged from the point of view of colonial features, serologic reaction, resistance to bacteriophage and cultural stability, the G forms manifest the expected characteristics of a distinct dissociant, and are therefore entitled to a place among other recognized cyclostages occurring in the development of the species. The distinctly recognized culture types thus become S, O, R and G, of which the G form is the least commonly seen in laboratory cultures. Although each of these can be easily recognized in dissociation studies as a distinct colonial and cultural entity in the species composite, their mutual relations, their origins and exact sequences, are not yet clear, except in a few cases. The main point that we desire to emphasize is not, however, disturbed by this circumstance: The G type culture with its attendant filtrable forms is established as a distinct and highly significant unit in the heterogeneity of culture forms constituting the Shiga and other species.

PROBLEM OF BACTERIAL LIFE CYCLES IN RELATION TO G FORMS

Another point of interest involved in the relation of the G forms to microbic dissociation is their possible interpretation with reference to bacterial life cycles. This constitutes an issue over which there is much variance of opinion. The existence of cell transitions regarded in the light of a "cycle" of development has long been supported by a feeble minority among bacteriologists, although it must be admitted that the chief notions involved have, as a rule, been indefinite and have found little acceptance among present workers. Indeed the majority of bacteriologists seem to develop an attitude of suspicion and defense whenever the term bacterial life cycle is mentioned. This is perhaps because the phenomenon is pictured as something difficult of comprehension and, above all, "complex." This situation, particularly with reference to the assumed "complexity," is indicated by the fact (readily observed in the writings of those opposed to the view) that one seldom sees employed the term "life cycle." It must needs be "complex life cycle." The notion of "complexity" has become firmly wedded to the notion of a life cycle, and is given voice to on all possible occasions as if to ward off any intending believers from a form of biologic heresy the fundamental issues of which may, as a matter of fact, be relatively simple. Indeed, there is no reason to doubt that, among the life cycles of microorganisms, some might be highly complex, while others might be very simple. It is only the latter of which we possess any very definite knowledge. There is involved here, however, a situation which, because of the great importance of the problems involved, seems to demand some further consideration.

In attempting to explain the phenomena of bacterial variability, which for many years has been an accepted matter, bacteriologists have had recourse to two chief views: one, that the observed variations are fortuitous and therefore disorderly; the other, that they are directive and therefore orderly, although the exact sequences have never been very clear. Those who support the first view adhere to the principles of Darwin or of de Vries, depending on whether the observed variations are regarded as transient or as "permanent" hereditary departures from the "normal" culture type. If the former, they are regarded as fortuitous or "impressed"; if the latter, they are regarded as actual mutants in the de Vries sense. Those who support the view of a directive variation regard the variations in question as the natural result of a distinctly directive culture development in which the cells, by virtue of some inner, guiding mechanism, pass from a simple, basic form through the distinct stages of ontogeny in their progress toward a predetermined end, the "adult" cell form, or the "Kulminante" of Enderlein. In a similar way, perhaps, any of the cells of a fungus filament hold in themselves the potentiality for the later formation of conidiophores and for the still later reproductive conidia. If bacterial cells in successive generations, instead of each becoming separate and going its independent way, could remain united for hundreds or thousands of generations, might we not be able to observe in this drawn-out and extensive "colony" certain cell differentiations analogous to some of those occurring in the higher fungi? Already we recognize something of the sort in long chains of streptococci, in which a considerable degree of cell differentiation can often be detected; also in such "giant colony" formations as those studied by Faith Hadley, working with the enterococcus (personal communication to the authors). Leaving aside all matters appertaining to sexual reproduction by bacteria, the possible existence of which still rests on very meager information, it seems that those bacteriologists who deny the existence of something equivalent to "seed forms" in bacteria and their gradual evolution into different cell types are assuming considerable responsibility. Yet the existence of such bodies has been, and is, denied by the great majority. In a sense, this situation is remarkable; for to anticipate that bacteria should possess some sort of "seed form" is the most logical deduction that could be made from present knowledge of forms most closely related to them. The point which to the present time has not received adequate demonstration is that these "seed forms" can be isolated, and that, in the course of time, they can transform into the original culture type. Proof of these facts is presented in this paper.

To the present, although sufficient evidence has been supplied to demonstrate a "directive" course in the appearance of bacterial variants, as one of us (P. H.) pointed out in a previous paper (1927), the distinctly cyclical aspects have, in most cases, been delineated only vaguely, and much has necessarily been left to the imagination. Many investigators who have, in recent time, abandoned the monomorphic conception—at least, to the extent of accepting the fact of common, more or less orderly occurring and stable variations, manifestly not of mutational significance—have progressed further only to be confronted with the question of whether these transformations are merely of a "back and forth" sort between two points of greater culture stability (such as the S and R types), or whether the line of transformation is really a "cyclical" affair.

Before considering this point or attempting to answer the definite question regarding the possible relation of the G type cultures to such a cyclical development in bacteria, it may be helpful to consider first what conditions would need to be fulfilled in order to justify the belief that a cyclical development of bacteria exists. It seems sufficient to suggest that the culture must pass progressively through three (as opposed to two) stages of development, and then return to the first cell type. Sexuality, it may be pointed out, has no necessary relation to this problem. As three points in space may determine a circle, so three stages of development, encountered successively or progressively and including a return to the first or basic stage, may be said to determine a life cycle. For this reason, it is not at present possible to picture a cycle involving merely the S and R culture types alone. And a cyclical phase becomes no clearer perhaps when we introduce the SR type, observable in many species and pictured clearly by Nungester in the case of *B. anthracis*.

But this difficulty does not appear to be present when the G form is introduced; for this culture type, in returning to the original form, passes through stages which, in our experience, are never encountered in the passage of the "normal" culture to the G form. The S or R to G transformation is extremely abrupt; the G to S (or R) very gradual. In other words, there are encountered colony types lying between G and S in the G to S transformation, which we have never observed occurring between S and G in the S to G transition. This indicates that the path of the G to S transformation includes culture forms of a third type that must be reached before the full S can be attained. Under these conditions, therefore, it appears that the culture passes progressively through at least three developmental stages; and this, we believe, is sufficient to justify the use of the term "cycle." The only instances, within our knowledge, in which a cycle has been worked out with any degree of completeness are, first, the case of *Azotobacter,* reported by Löhnis in 1923; second, the case of *B. anthracis,* reported by Haag in 1927, and, third, the case of *B. mycoides* reported by Oesterle and Stahl in 1930. Tunnicliff's *Bact. gonidioformans* probably holds the same significance, and there are a few more somewhat obscure references in the literature.

To summarize the last few pages—we are of the opinion that the existence of a cycle of development embracing the mature cell, the seed form and certain intermediate types (G forms) lying between the seed form and the adult stage is not only indicated by analogy, but also demonstrated in fact by the chain of culture elements described in this paper.[26] It is also highly probable that some of the so-called seed forms are identical with the filtrable elements associated with the G type cultures, and that these elements represent the bacterial microgonidia, some of which are probably invisible by ordinary methods of microscopic examination. We are, further, inclined to the view that the period of apparent noncultivability of the filtrable bodies may be related to the clearly recognized "resting stage" characteristic of all the gonidial forms that have been studied in the higher fungi. In our opinion, these observations and conclusions thus serve to bring the G type cultures into definite relation with microbic dissociation and, pari pasu, with bacterial cyclogeny. It is within this category of phenomena that the filtrable bodies, like the bacteriophagic corpuscles, will, we believe, find their final explanation.

RELATION OF G FORMS TO "YOUTH TO AGE" VARIATIONS OF HENRICI

The discovery and interpretation of the G forms of culture have a bearing on certain recent attempts to explain the nature and significance of bacterial variations on grounds quite apart from the older views (fluctuating variations, mutations), and equally apart from the more recent views of dissociative behavior. Henrici, in particular, undertook to fit these variations into a scheme embracing the morphologic changes that occur in the life of the culture between the "embryonic" and the "old age," or senescent, forms. These have sometimes been termed the "youth to age" variations. According to Henrici's view, the chief morphologic variations seen among bacterial cells are dependent on their age, and are correlated with their rate of growth. He presented many sketches of these changes made hour by hour in the life of the cells. His "old age," or senescent, forms he believed are identical with the "degraded" or "involution forms" long known to bacteriologists. All of these variations he brought together under the heading of "cyto-morphosis."

Henrici's general scheme involves issues which he held are somewhat opposed to the older monomorphic conceptions of bacterial stability, but we find it difficult to ascertain where the difference in point

---

26. By this statement we do not mean to imply that the cyclical or phasic aspects of bacterial development are necessarily limited to this phenomenon.

of view really lies, since most of the monomorphists have accepted a degree of variability in cell form at least as great as that presented by Henrici. Aside from the circumstance that he added to the category of variants those which he termed the "embryonic" forms, and suspected the existence of different physiologic capacities in the different age groups, he seems to have departed but a slight distance from the earlier conceptions of the variation problem. He expressed the belief, moreover, that his hypothesis is not incompatible with some of the more radical views of pleomorphism and life cycles, while at the same time not accepting the significance attached to the supporting observations by Mellon (1920) and others.

As a result of our earlier studies on microbic dissociation, and especially with our present knowledge of the G forms, we find ourselves unable to accept the view of Henrici in any of its major features, although we freely admit the truth of his statement that certain morphologic characters are associated with the age of the cells and correlated with the rate of growth. While admitting that he demonstrated interesting morphologic changes, we also believe that the essential features of microbic variation, as supporting the broader view of pleomorphism, lie for the most part outside the field of study that he covered in his laborious and carefully prepared work.

According to our view, the chief failure of Henrici's hypothesis lies in the circumstance that he failed to take into sympathetic consideration the fact that the variability phenomena in bacteria are to be found in two quite different categories, both of which were emphasized in the work of Fuhrmann and of Enderlein. One of these, and the only one that Henrici seems to have considered seriously, comprises the variations which are manifested by the cells of the growing culture over a brief span in its life history, perhaps only between successive cell divisions, while the organisms are multiplying exclusively by fission. These variations are interesting, and some of them (particularly those included in his senescent forms) are important, but the majority we regard as possessing slight significance from the point of view of the broader problems of variability and physiologic differentiation.

The second category of variation, already mentioned, comprises the more significant variations which occur over a longer period of time—days or weeks or, perhaps, months; changes for the detection of which not one cell or one culture tube, but many successive cultures, must be followed over a long period. This is the "fortschreitenden Entwicklung" of Enderlein. These changes are progressive and appear to be due mainly to the assumption of modes of reproduction which, to say the least, are different from the conventional simple fission, and which are entered on at different times in the history of the cultures, dependent in part on the nature of the environment.

With this second category of variation phenomena, Henrici concerned himself little, if at all. In his attempt to portray the range and significance of morphologic variation, he studied (aside from his "senescent" stages) only the more conventional forms of growth, those that grow in the conventional patterns on the conventional mediums. The few stages of growth that were of significance among the forms studied, he dismissed as dead or dying forms, and their ultimate history was not followed. In other words, he seems to us to have sacrificed a wealth of variational material in order to avoid the pitfalls of dissociation and its cyclical connotations.

With reference to this point, the long tangled filaments of the R type of *B. acidophilus,* as studied and recently reported by Faith Hadley (1930, b), and the analogous growth forms of *B. welchii,* as studied by Roe in this laboratory, represent distinct morphologic variants for these species, and variants that remain stabilized for weeks or months. These forms in pure cultures may grow rapidly or slowly, and, among themselves, they may be longer or shorter, broader or narrower; but they represent only the fungoid stage of the species. Moreover, these forms have no significant relation to the age of the independent culture. The minute granular or coccoid bodies of the Shiga bacillus or of *B. typhosus,* or of the diphtheria bacillus, as studied by Miss Richardson in this laboratory in connection with the filtrable forms, are distinct morphologic variants of these species, but they also may remain stabilized for a considerable time. Moreover, they have no necessary or significant relation to the age of the culture as measured by hours. The diphtheroid rod stage of a streptococcus culture (Koch and Mellon) is a distinct morphologic variant which may remain stabilized for a long period, and may reveal within itself perhaps as great a variation as that pictured by Henrici for any of his rod forms. But it is only one stage in the ontogeny of the streptococcus species, and has no relation to the age of the cells of the culture as measured by hours.

In all of these cases, however, there is a distinct relation between cell morphology and the stage of the organism in its ontogenic development, from the earliest "embryonic" stages to the adult, reproducing form. In other words, the observational material that Henrici presented is limited, for the most part, to growth phases which comprise but a fraction of the whole complicated picture of bacterial variation within the ontogeny of the species. We believe, moreover, that the methods employed by him in his approach to his difficult problem are not adequate to obtain the anticipated results. In order to study the really significant variations among bacteria, one must first dissociate his cultures into their component cyclostages, and then develop these cyclostages in approximately pure lines.

When this is accomplished, as it easily may be by suitable means, one can readily observe that cell types which seem to correspond well with some of Henrici's "age groups" can be stabilized to such an extent that the independent colonies of the respective variants reveal almost pure populations of their own morphologic type. The freshly isolated G colonies are made up almost exclusively of the coccus or granular bodies the larger of which Henrici perhaps observed in some cultures. The S type colonies are made up chiefly of rods of medium length. similar to the majority of the adult cells. In pure S cultures there is little variability. The SR colonies involve a mixture of medium and long forms, while the well purified and stabilized R colonies comprise longer rods and filamentous structures which Henrici failed to picture among his variants. Each one of these morphologic types may remain stabilized over considerable periods of time—days, weeks or months—showing practically no intergradations with the other forms. Planting the granular or coccus forms in a fresh medium yields, from the first hour of growth, only the coccoid bodies, although after days or weeks of transfer these forms may elongate. Planting the medium-sized rods from pure S colonies into fresh mediums yields only the medium rods, and planting the long filaments from pure R colonies yields only long filaments, although after days or weeks, coccoid bodies may begin to appear in, or outside of, the filaments.

We do not wish to say that some slight variations in shape or size, or in the length-breadth index, do not occur among these morphologic types. But we state that no strict relation exists between the chief morphologic types mentioned and the age of the cultures in hours or even in days. The embryonic form persists even though the cells of the culture are old; the medium-sized rods persist whether the cells be young or old, and the long filaments persist even though the cells of the culture are young. The size relations depend on the stage in ontogeny, not on the age of the culture elements within a single tube. These observations may be made by anyone who takes the trouble to dissociate his cultures into their essential cyclostages and then works with the well stabilized types as indicated by the characteristic colony form on a favorable medium.

Henrici spoke frequently regarding the cells that he characterized as "physiologically young." What did he mean by this term? Obviously this characterization applies to cells that have recently divided or to cells that are found in the earliest growth stages in a fresh medium. They are physiologically young with reference to the span of life in a single tube of culture. Henrici stated (p. 148 of the book cited in the bibliography) :

There are scattered references in the literature to differences in degree of virulence, of antigenic value, of susceptibility to antibody action, or differences in fermentation power, between young and old cultures; but in practically no cases have these variations in physiological properties been definitely correlated with phases of growth. It would seem that this would offer an extremely attractive field for investigation, especially from the standpoint of infection and immunity. Such investigations, however, are difficult to carry out, because when the cells of bacteria are physiologically young, they must necessarily be few in number in proportion to the volume of the medium, so few that they are not suitable for most types of experiment. Every attempt to concentrate them, particularly to remove them from their medium, immediately causes them to rapidly age, and they are no longer physiologically young but mature cells. For this reason I do not believe that very much work has so far been done with physiologically young bacteria.

Against this point of view, which is erroneous, because based essentially on monomorphic conceptions, we contend that physiologic youth of bacterial cells has nothing to do with the period of existence between cell divisions or with the early growth stages in a fresh culture medium that is being observed over a period of hours. We prefer to believe that true physiologic youth is, on the contrary, dependent on the stage of advancement of the individual cell through that range of culture existence lying between the development from the bacterial gonidium and the assumption of the adult or reproductively mature form. This range comprises bacterial ontogeny. One is just as likely to find cells characterized in this manner in cultures days or weeks old as in cultures a few hours old. Physiologic youth, if we give this term its customary biologic meaning, is related to a stage in the ontogeny of the individual, and cannot be applied with exactness to the stages of bacterial development within a single tube culture. This is because the individual bacterial cell does not represent the entire species; it can represent only one cyclostage of the species. The complete species is made up of the summation of its cyclostages, which, in turn, comprises the ontogeny. This is true for any animal or plant organism, and is no less true for bacteria, although in this group of living things the separation of the successively formed cells renders the interpretation more difficult.

Guided by this interpretation, we further conclude that true physiologic youth in bacterial cells is found exclusively in those coccus-like bodies (or related elements) that represent the youngest of the G type cultures. They alone are in the earliest period of ontogenic development; and we have noted how different are the physiologic, serologic and immunologic reactions of these cultures, compared with those made up of more mature cells.

Furthermore, it seems to us that physiologic age is not necessarily an attribute of the oldest cells in a single tube culture, but is found only in those cells occurring during the later stages of ontogeny. To sum up the matter, it is possible to apply the terms "youth" and "age" either to the cells in a single tube culture or to those cells that appear successively in the unfolding of the ontogeny of the species. In the strict biologic sense, however, these terms are applicable only to the definite stages of ontogenic development. Moreover, it is these stages (cyclostages) only that are concerned with the problems of pleomorphism and bacterial variation in the larger sense.

So far as Henrici's so-called embryonic stages in variation are concerned, it is therefore clear that, if one wishes to discover this form in any organism, it is scarcely sufficient to start with a culture isolated from a stage in cyclogeny that is perhaps already well advanced. The embryonic stages of a fungus or a yeast are those cells that are just emerging from the spore or conidium. In a similar manner, what may be termed the true embryonic stages in the ontogeny of a bacterial species are represented by those bodies that first arise from the bacterial gonidium, and then begin their slow and gradual ascent toward the adult cell type. With these, Henrici's account does not concern itself.

Likewise, the adult or mature cell is in reality that protoplasmic unit the germplasm of which has carried through its development, which has passed the period of vegetative cell multiplication, and which, after a long period, has attained reproductive maturity. This stage, we believe, is often represented by the pure R type culture. At this point, which represents the actual peak of ontogeny, the gonidia, or other physiologically equivalent reproducing bodies, are formed and liberated; and, with their liberation, the old cell may be said to "die," since its function in life has been completed. But it dies in the process of giving birth, for the germplasm passes on to new and quite different morphologic and physiologic units.

This interpretation of variation phenomena, though still lacking in some of its details, affords a picture of bacterial development very different from what Henrici painted. It enables us to understand that knowledge regarding the extent and character of variability in a bacterial species cannot be secured from even the most minute and mathematical study of the cells occurring in a single phase (cyclostage) in its life history. A few hours' study at a certain time, regardless of the number of cells measured and statistically tabulated, does not inform us in what guise the culture appeared several hundred generations earlier, nor how it may appear some hundreds of generations hence. Bacterial variability in its most significant form is not adequately expressed within the brief period of a few cell generations, nor necessarily within a few tube generations. It is most commonly tardy in its development and has a tendency to be cumulative over long periods of time. Before the wide range of morphologic variability or of physiologic capacity in any bacterial species can be fully understood, the development of that species must be followed, cell after cell and tube after tube, from the "seed" to the adult form. This gives, in the only manner possible, a conception of the ontogeny of the species. And, if one does this, one will find that the outlines of bacterial variation presented by Henrici are of too limited and fragmentary a nature to yield more than an imperfect view of the great range of these phenomena and their significance in the life of the species.

SIGNIFICANCE OF G TYPE CULTURES AND OF ASSOCIATED FILTRABLE FORMS IN COMMON LABORATORY PROCEDURES

Each day it becomes more apparent that a recognition of the essential facts of microbic dissociation, especially those dealing with the S, O and R cultures and their characteristic antigens, is playing an increasingly significant rôle in applied bacteriology. It appears probable, moreover, that this significance will be increased and extended into other lines by the recognition of the G type culture and its associated, filtrable elements, which cannot fail to enter into the prosecution of bacteriologic laboratory work of diverse nature. Some of these possibilities may be considered.

. When an old agar slant culture, on attempted transfer to fresh agar or broth, fails to register growth, one is likely to conclude that the culture is dead. Cultures of the Shiga bacillus are notably difficult to keep "alive" for a long period on agar slants. The nature of our results with some of these "dead" slant agar cultures raises the question : When is a culture actually dead, and when has it merely been transformed into a cyclostage in which it fails to reveal the signs of vitality that we have become accustomed to observe in living cultures? It is well recognized that, in old cultures, it is usually impossible to discover organisms which, on staining, reveal the typical morphology and staining reactions of the original, living form. The living organism is there, as successful cultivation often demonstrates, but it seems to be present in a different form from any with which one is well acquainted.

Enderlein, in his comprehensive study of bacterial cyclogeny in 1925, made an interesting allusion to a phenomenon probably identical with the one being considered. For the revival of old, and apparently dead, cultures on solid mediums, he recommended the following procedure : Make a transfer from the culture, not to agar, but to broth, and place the tube for some hours in medium sunlight. After from six to seven hours seed the broth back to agar, after which growth is likely to start on the solid medium. In explaining what happens in the old agar culture, Enderlein presented the following interpretation : With the increase in the unfavorable conditions in the old agar culture, the organisms begin to enter a new cyclostage, characterized by the coccus and coccoid elements, and representing the gonidial forms. In time, the culture may be made up largely or entirely of such bodies. If conditions in the medium become favorable, these bodies may, by division, return to the higher, or original, cyclostage. If the adverse conditions continue, however, or under the continued influence of high temperature, or of light, the gonidial bodies use up their small amount of food-reserve and, by further division, enter the more minute "gonite" stage. These minute coccus forms, practically destitute of cytoplasm, are no longer capable of growth on solid culture mediums. In broth, however, they may, within from five to seven hours, give rise by further division to the sexual elements, the "oites" and the "spermites." Between these bodies copulation occurs in the liquid medium (but rarely on solid) and the freshly fertilized cell begins its evolution toward the higher cell type, thus entering the new cyclode. The organism is now again susceptible to cultivation on solid mediums. Of these various forms— gonidia, gonites, oites and spermites—Enderlein presented a description, giving in detail their form, size and nuclear structure. In these changes he recognized the existence of a definite sexual cycle, most clearly observed in the cholera vibrio.

Regarding the details of this process, the recognition of which requires a concise and somewhat difficult cytologic technic, and regarding the exact interpretation of events reported by Enderlein, we are not in a position to give either a confirmation or a denial. It is only fair to say, however, that our results, dealing with a cyclostage that, for a time at least, is not directly cultivable in colony form on agar plates, but that becomes cultivable on agar after it has been sufficiently incubated in broth (or carried in series over moist agar plates), find a close agreement with some of Enderlein's observations.

These observations thus possess some suggestiveness, we believe, in relation to the probable nature of the events occurring in old and "dying" or "dead" bacterial cultures. They also indicate the nature of the criteria that must be applied before it may be safely concluded that a culture is actually destitute of living elements. In addition, they afford a means that may, under certain conditions, be effective in recovering an apparently deceased strain; also a means that may reveal the presence of living elements in blood, tissues or filtrates judged sterile when the customary methods of testing are employed. In view of the practical significance of these considerations, it is perhaps desirable to present in concise form the methods that may be applicable in such cases, so far as they can be formulated from our own experiments and those reported by Hauduroy (1927, a), who was the first to employ the serial agar plate method. The following considerations refer primarily to cultures and culture filtrates, but we have found that the same technic may be applied with equal effectiveness to the examination of the blood and of pleural and peritoneal exudates in rabbits and guinea-pigs, for living elements.

Methods of Detecting the Filtrable Forms.—The first consideration concerns the appearance of visible growth in samples of the original Berkefeld filtrate. Our method has usually been to divide the fresh filtrate into small samples of from 5 to 8 cc. each, to incubate them for several days at 37 C., and then to keep them at room temperature for continued observation. Although a faint opalescence in the broth, after a period of from three to thirty days, may reveal the presence of living forms, the more common evidence in Shiga filtrates is the formation of a slight, somewhat viscous or viscogranular sediment, the amount of which increases with time. The amount may, at first, be so slight as not to attract notice, unless one is anticipating its appearance. It resembles closely the traces of sediment observed in aged bacteriophage filtrates, from which cultures resembling the G type can usually be obtained. It must, however, be differentiated from filter sediments and other broth sediments, and must be checked by microscopic examination and by the appearance of sterile control tubes incubated under similar conditions and for the same length of time. If the sediment appears, or even if it does not appear (which is likely to be the case with fresh filtrates), the test must be carried further as follows. The continued tests, however, should not be made until the filtrate is from five to ten days old. These tests involve the inoculation of broth tubes and of agar plates in series. For reasons that we do not yet understand, litmus lactose agar seems preferable, as first pointed out by Hauduroy (1927, a), for the filtrable forms of a number of organisms. For obvious reasons, we believe that mediums containing blood, ascitic fluid and living tissues should be avoided if possible. If used, several control plates should be carried under the same conditions of experiment, sterile broth replacing the filtrate for plate inoculation.

The inoculation of broth tubes should be made by transferring a drop or more of the sediment, if present, taken with a pipet from the bottom of the tube. If the filtrate does not contain a sediment, 5 drops or more of the filtrate itself will usually suffice.[27] The broth tubes should be examined after some days to ascertain the presence of opalescence or of sediment. The sediment, if found, should be examined in films stained by the Giemsa or Romanowsky methods. From this time, continued transplants may be made in the same manner. In the course of such transplants the culture elements can be recognized sooner or later, although a careful differentiation from stain deposits and other artefacts is required. After a long series of transfers, the broth may begin to reveal a faint homogeneous clouding, which increases in density with continued transfer. By this time the culture may grow on agar, depending on how rapidly the regeneration from the filtrable elements has occurred. As a rule, however, it may be anticipated that many transfers may be required before the latter event occurs. If the seeding of an apparently fair growth in broth to an agar plate does not reveal growth after from forty-eight to ninety-six hours when the plate is examined under the no. 3 lens, serial plate transfers must be instituted, after the fashion detailed in the following paragraph.

The presence of living elements in the filtrates can often be ascertained more quickly by broth-to-agar cultivation, although the rapidity with which the G colonies appear on agar plates depends on, among other factors, the age of the filtrate when the agar plate is streaked. If a fresh Berkefeld N or W filtrate is streaked when less than from 5 to 10 days old, it is less likely that growth, in the form of visible G colonies, will appear on the first plate. In this event, a seeding must be made from the first plate to a second, from the second to a third and, in the case of continued failure of growth within three days, even through still further transfers. It happens in a considerable number of cases,

---

27. From the statements made above, it might appear that the use of a considerable volume of filtrate for the inoculation of the agar plates and of a considerable volume of broth "washings" for the plate to plate transfers in series indicates the presence of only a small number of cultivable elements in the original filtrates. This is not the case. It appears to us that the chief reason for the use of considerable volumes of filtrate or broth for inoculation of the plates is to afford sufficient moisture on the agar surface. In this respect, the work of Garrod on the cultivation of filter-passing anaerobes isolated from the upper part of the respiratory tract has a significant bearing on our results.

however (in Shiga and related cultures), that if the sediment or the filtrate itself is streaked after from five to ten days, a very faint growth in the form of visible G colonies will appear on the first plate after from forty-eight to ninety-six hours' incubation. They may at first, however, be detectable only by the aid of a hand lens, and sometimes they are microscopic in size. Satisfactory amounts for seeding the agar plates are: 1 or 2 drops of sediment from the "grown" filtrates, or from 5 to 15 drops of fresh filtrate. Hauduroy (1927, a) used even larger amounts and recommended permitting the filtrate to stand on the agar until it has been absorbed. As a rule, however, we have found no advantage in using more than from 10 to 15 drops ((0.3 to 0.5 cc.).

In examining the stained films made from either the broth sediments or from the growth on plates, it must not be anticipated that the morphology of the cellular elements revealed will bear any close resemblance to those of the original culture. Moreover, a marked variability in the morphology of the elements that first appear should always be expected. The final test of the identity of the culture so isolated is the enforcement of its "reversion" to the culture of origin or to a related and recognizable cyclostage identifying the species. This reversion may occur within a few weeks, or it may require months. As a rule, we have not observed the rapid reversions (taking place in the course of eight serial plates) reported by Hauduroy (1927, a) for the filtrable stage of the streptococcus. In fact, in some of our cultures, a return to the original type has not occurred to the time of writing. How long such G type cultures may remain thus stabilized we do not yet know. But we believe that their ultimate transformation to the original culture will occur in all cases, just as we have some reason to believe that it occurs in all R type cultures, which also often manifest marked stability. Hauduroy also noted some very stable forms. There probably exist, however, some means of which we have no present knowledge of forcing more rapidly the retransformation of both types of culture.

Another matter related to the unsuspected presence of the G forms concerns the actual purity of bacterial cultures isolated from colonies on plates. In selecting a colony for subculture, it is seldom the practice of bacteriologists to examine the colony otherwise than by the unaided eye, to be assured that it is discrete. Sometimes a hand lens is employed. but seldom the lens of a microscope, unless a low-power binocular. From our experience, we cannot advocate too strongly the necessity of scrutinizing every colony selected for subculture, at least with a no. 3 objective, in order to discover, in it or very near it, minute colonies of the G form of the same, or perhaps another, species. If it happens to be of the same species, of course it does not greatly matter; if of a different species, the circumstance naturally possesses significance. The G colonies may have developed within the larger colony, or the larger colony—especially if it is of the spreading R type—may have overgrown a smaller colony.   Beyond this, however, we also have reason to suspect that an apparently pure colony may have associated within it living G forms that are not detectable by microscopic means.

A further aspect of bacterial study on which we believe our results may also have a bearing relates to those experiments performed by various investigators, particularly by Frobisher and Brown, in which two different species of bacteria, for one purpose or another, are grown together in the same culture tube, or in which one organism is grown in the filtrates of another.   In most of the experiments of this sort reported it has been the aim of the investigator to demonstrate that one of the associated organisms becomes modified as a result of contact with the other, or with some of its products, such as those present in filtrates.   When, in such tests, the two organisms concerned represent species widely separated from each other, as *B. melitensis* and *B. anthracis,* used in some of the studies reported by Burnet (1925) on the phenomenon of "entrainement," conclusions regarding the fact of culture modification are probably justified.   When, however, as in the work reported by Frobisher and Brown, two strains of streptococcus, one toxigenic and the other nontoxigenic, are associated for purposes of growth in the same culture tube, any conclusions regarding the modification of one strain by the other are, to say the least, based on distinct uncertainties.   The reason is that, if there are filtrable, and for a time noncultivable, stages in the life history of a bacterial culture (it now seems well established that filtrable forms of the streptococcus exist), there may be a doubt whether the organism reisolated after its alleged modification has actually been modified, or whether it has been confused with the organisms developing out of the filtrable stage of the associated strain.   In associations between such closely related organisms as toxigenic and nontoxigenic streptococci, as employed by Frobisher and Brown in their work, the conclusion favoring culture modification must, for the time, remain in doubt.   And this view of the matter is not modified by the outcome of Frobisher's late work on filtration, in which he was unable to demonstrate the filtrability of any of the organisms studied, and from which he concluded that the positive results obtained in all previously reported instances were likely due to faulty candles or to contaminating bacteria, of which Frobisher reported astonishing numbers in his own filtrates.

In a similar way, we are led to doubt the reliability of all methods of single cell isolation employing the selection of a single organism from a liquid medium. The minute and perhaps invisible corpuscle representing the filtrable stage of the culture may, at any time, be present in the medium and in the micropipet. The method of often repeated platings on a favorable medium, accompanied by careful scrutiny of the selected colonies when fully developed, and by means of the microscope, we believe offers a safer means of isolating pure line strains. But it is also apparent that even this method is not above reproach.

Criteria for Filtrability of Bacterial Cultures.—Continuing our reference to methods of detecting the filtrable forms of bacteria, we desire to present what we believe to be some useful criteria for ascertaining whether a given culture is filtrable—requiring, of course, in most cases, the presence of the special filtrable forms characteristic of the G type cultures. Here we limit our consideration to filtrates from Berkfeld N and W candles or to their Chamberland or other equivalents.

In general, we conclude that a culture is not filtrable (in the sense in which this term is used in this paper) if the following circumstances obtain:

1. If a visible clouding or other indication of growth appears in the filtrate, or in subcultures from it, within a period of from twenty-four to forty-eight hours from the moment of filtration.

2. If a slight turbidity, opalescence or slight viscogranular sediment does not appear in the tubes of filtrate sealed in ampules and stored for from six months to one year.

3. If the colonies that first appear on plates seeded with the filtrate are of "normal" size and appearance, and particularly if their diameter, after ninety-six hours of growth, exceeds 0.2 mm.

4. If the organisms that arise in the filtrate, or in subcultures, manifest a resemblance to the cells of the parent strain in morphology or cell grouping. Cultures the "normal" form of which is a micrococcus or a streptococcus are excepted.

5. If the serologic and biochemical features of the recovered culture agree with those of the parent culture, and if the recovered culture is still susceptible to the homologous bacteriophage.

In general, we conclude that a culture is filtrable (in the sense indicated in this paper) if the following circumstances obtain with reference to the recovered form of growth:

1. If visible signs of culture development appear in the filtrate, or in subcultures from it, only after a considerable delay, perhaps after from six to twelve days, or even after a longer period in the case of some species.

2. If the colonies on agar plates are very slow in appearing, perhaps after from twenty-four hours to eight days, and if the first colonies to arise are of minute size, having an average diameter of 0.2 mm. or less at the end of ninety-six hours on a favorable medium present in sufficient depth in the plate.

3. If the morphologic type of the organisms present in the filtrate, or in sub-cultures or colonies, is composed mainly of minute coccus forms, granular bodies and delicate skeins or filaments with a suggestion of diplococcus or streptococcus arrangement (Giemsa staining) ; also occasional coccobacilli.

4. If, in organisms the "normal" form of which is acid-fast or gram-positive, the staining reaction of the recovered culture elements is temporarily reversed or variable.

5. If the biochemical and serologic reactions of the recovered cultures are different from those of the parent culture, and if virulence, toxicity and susceptibility to the homologous bacteriophage are altered.

The statements here presented are intended merely as rough generalizations which we venture to formulate as a result of our observations on the filtrable forms that have been produced in this laboratory for a considerable number of bacterial species. They are, moreover, generalizations that are in strict accordance with the observations and conclusions of Hauduroy (1927, a), who studied the virus forms of several pathogenic species.

Outside of these instances, however, we reserve the possibility that certain micro-organisms exist the "normal" form of which is such that it readily passes the filter candles. In this group might be placed *Bact. pneumosintes, Asterococcus mycoides* and perhaps a few less well known organisms.

RELATION OF G TYPE CULTURES AND OF ASSOCIATED FILTRABLE FORMS TO BACTERIOLYSIS

In relation to the frequent disappearances of the S type culture just preceding, or accompanying, the first appearance of the G forms, we desire to consider briefly the possible nature of bacterial autolysis.[28]

It is perhaps safe to say that, to the majority of bacteriologists, the term "bacteriolysis" carries a meaning synonymous with the "granulation" and consequent death of the bacterial cells. It is believed to occur naturally in old cultures, but has been studied particularly in the vibrio of cholera under the influence of immune serum; also to some extent in *B. typhosus*. The problem is also closely related to the com-

---

28. By this term we do not refer exclusively to "transmissible bacterial autolysis" or bacteriophagic lysis, but to the phenomenon of alleged bacteriolysis, recognized much earlier, particularly for the cholera vibrio.

monly alleged "germicidal" influence of immune serum and some normal serums. Of special interest in this connection are the older studies of Pfeiffer, Radziewski, Eisenberg and others. Notwithstanding the common acceptance of the view that bacteriolysis is accompanied by the death of the cells, the investigators mentioned have presented additional evidence which demonstrates that the autolysis determined by immune serums or by some "toxic" normal serums does not always cause the death of the cells—at least, of all the cells—but a transformation of the culture type. This was manifested on agar plates when they were examined from three to six days after they had received the culture and the serum. Often there occurred a fine sprinkling of minute colonies containing minute organisms which, by slow steps, again grew up to the normal culture form. The failure of the original culture forms to appear was regarded as sufficient evidence that the greater portion of the cells had been "killed" by the action of the serum.

What is the actual nature of bacteriolysis? There has been a tendency, from the beginning, to attribute its visible reactions (granulation of the organisms and apparent death) to "enzyme action"; and that is probably the reason that so many investigators have favored the interpretation of bacteriophagic lysis as representing some sort of self-digestion. Of course, it cannot be doubted that bacterial cells, like any other animal or plant cell, may, under certain conditions, become susceptible to autolytic changes of this sort. On the other hand, it may well be doubted that the cells ever succumb to such a process until they are already physiologically "dead." To attribute the death of an organism to a self-destruction due to enzymes arising within it or impressed on it from without is taking great liberties with what we know of cell physiology. There is good reason why such a reaction should be regarded as highly "pathologic," even as the proponents of this notion admit; indeed, it would be regarded as highly extraordinary.

Moreover, the "granulations" that appear in such "dying" cells are open to an interpretation quite different from that of protein disintegration, for similar granulating tendencies can be observed in many cells at a certain stage of the dissociative reaction. On the other hand, it is of course true that bacterial cells die and disintegrate. It thus seems reasonable to conclude that in the phenomenon of bacteriolysis there may be present two quite different cell reactions, one an indication of degeneration, the other representing a stage in cell transition into smaller units. The "lysis" of a culture by the bacteriophage has a bearing on this point. The "normal" cells disappear from view, and it is easy to conclude that they have been destroyed by lysis. Although in such cases the cell forms perceived earlier certainly disappear, we have good reason to believe, from experiments performed in this laboratory, that the living germplasm of the culture is never destroyed in this reaction or by this means. Indeed, it appears highly doubtful that even the most voracious strain of bacteriophage ever actually "destroyed" a bacterial cell. One of us (P. H.) has pointed out repeatedly that the true function of the bacteriophage is not to destroy cells, but to transform them. Among the products of transformation are minute, filtrable bodies which eventually are able to regenerate into cultures that we are unable to distinguish from the G type; and from these the culture transforms further into the original form.

What has just been said regarding bacteriophagic lysis we are inclined to believe may also be true of the common bacteriolysis. The culture seems to disappear; the usual colony forms are not found on the plates. If these results are obtained through the use of an immune serum (for instance, cholera antiserum), we say that they are due to its "lytic" or "germicidal" power. We doubt whether such serums possess appreciable germicidal influence. We already have good reason to believe that in many cases, at least, the bacteria have been transformed into a new cell type which, for a time at least, possesses slight growth energy and which is unable to register on plates in the usual colony picture.

Indeed, there is reason to believe that much that we have learned regarding bacteriolysis, whether of the "common" or of the "transmissible" type, must be unlearned. The same is true of the alleged germicidal action of immune, as well as of some normal, serums. At present these points cannot be definitely proved from the evidence in hand; moreover, the problem is somewhat aside from our main subject. In it, however, are involved issues of considerable significance for immunology, and it is fortunately open to experimental attack. In such a study we are confident that the filtrable forms of bacteria and also the G type cultures will play an important part.

UNICITY OR DUALITY OF FILTRABLE FORMS; POSSIBLE RELATION TO BACTERIOPHAGIC CORPUSCLE

Throughout this experimental report we have thus far spoken of the filtrable form of the Shiga bacillus as if it comprised a single type of element. Of the unicity of these elements we are not, however, completely assured. On the other hand, definite evidence of a duality of

form is by no means clear. The slight indications that we have perceived are perhaps unimportant, but, in view of certain related matters,
it seems desirable to add a word on this aspect of the problem.

Only twice did we make observations that seemed to make questionable the view of unicity. These exceptional cases concern the discovery of what we have termed the "rough G type" colonies. These
were colonies of the same size as the smooth G colonies, but characterized by a roughened or corrugated surface and irregularly serrated
borders. In a few cases, on microscopic examination of the rough G
colonies having a diameter of from 0.2 to 0.3 mm., the roughness
appeared to be due to the presence of still more minute secondary
colonies arising in the minute mother G colonies. In some cases, the
secondaries were as distinct, microscopically, as the G secondaries in
the mother S colonies (*B. typhosus*). How general this may be we
are not prepared to say. It merely indicates that even in the minute
colonies of the G type, still smaller, secondary colonies may exist. This
observation confirms that made by Haag (1927) on the colonies produced by the gonidial culture types of *B. anthracis*.

In the two cases mentioned, the origin and destiny of the rough
G type colonies was of interest. They arose from cultures that were
of the rough form of the Shiga bacillus, in one case under the influence
of lithium chloride, in the other case under the influence of pancreatin.
One of the rough G forms, after about sixty passages through plain
infusion broth, reverted to an R type Shiga culture which was typical
except for the circumstance that the colonies were slightly smaller than
usual. In the second case, the reversion of the rough G form was not
observed.

What the significance of these observations may be we do not
know. Perhaps there is none; but we have been especially tempted to
record them in view of the report by Haag of the finding of two distinct types of gonidial bodies in cultures of *B. anthracis*. Each of these
was accompanied by a distinct colony appearance—one round and
smooth, the other irregular and rough.

While we are dealing with these peculiarities of the G forms, we
also wish to mention another curious point, the significance of which
will doubtless sometime be learned. This relates to a peculiar duality
observable in certain morphologic bacterial elements existing within the
same species and strain. We can detect, for example, the most striking
duality in the two fundamental bacterial colony types, the S and the R
—the former small, round, regular, smooth, opaque and whitish, the

latter large, irregular, translucent, rough and bluish. Much of the vegetative life of any bacterial culture seems to be spent in oscillations between these two cyclostages, although long periods of stabilization in either one may occur. Again, we can note the dual system comprising the complete bacteriophage, as first clearly indicated in the work of Gratia, and as one of us (P. H.) independently (1928) and with Eugenia Dabney (1928) has already pointed out. Here we detect the alpha, as opposed to the beta, lytic units, each associated with a lytic fraction possessing its characteristic mode of action on sensitive culture, its especially appropriate substratum, its own type of secondary, resistant culture (following lysis) and its own peculiar antigenic structure. In the third place, we now catch a suggestion of the existence of a duality in the nature of the gonidial bodies which, as we believe, are either identical with, or closely related to, the filtrable elements of the G type cultures. At present it is idle to speculate regarding the possible significance of these dual systems, easily discoverable in the dissociative reactions of bacteria. They are, however, worthy of passing record.

These considerations lead, finally, to another subject which possesses some interest in view of the theory of bacteriophage action (homogamic theory) earlier proposed by one of us (P. H., 1928). It may be recalled that according to this view, the bacteriophage was postulated as a filtrable cyclostage in the life history of the culture concerned or in the life history of a closely related species. Assuming, for the moment, that this represents the facts of the case, the important question arises—What might be the relation between the filtrable elements preceding or associated with the G type cultures and the equally filtrable bacteriophagic corpuscles? [29] There may be none. But the

---

29. It will be apparent to the reader who has followed critically the earlier publications of one of us (P. H.) that he has been attempting to lead up to this crucial point. In 1927, the chief phenomena of microbic dissociation were reviewed and analyzed. In 1928, the chief facts relating to bacteriophage action were reviewed and analyzed, and correlations made with the dissociative reaction. In 1928, also, with Klimek, it was pointed out that the bacteriophage could be produced artificially in "normal" cultures as a result of forcing on them the dissociative reaction. In the present study, we demonstrate that the filtrable forms of bacteria can be produced artificially by enforcing the dissociative reaction on the mother culture. The general problem in the final study of this series thus becomes defined, namely, the possible relation existing between the bacteriophagic corpuscles and the filtrable bacterial forms which precede, and also coexist with, the chief elements of the G type cultures.

possibility has points of special interest. There can be no doubt that the bacteriophage corpuscles and some of the filtrable units associated with the G type cultures, or immediately preceding them in the culture tube, share several features in common.

First, they cannot be demonstrated in every "normal" culture or at all times in the same culture. As first suggested by Otto and Munter for the bacteriophage, and as further demonstrated by two of us (P. H. and J. K.), the first appearance of the bacteriophagic corpuscles is dependent on the occurrence of certain "Momenten" in the life of the cultures; moreover, these moments are often definitely timed, and in such a manner that they are correlated with the attainment of a certain stage of development, reached in a certain tube, in a series of broth passages. Regarding the moment of appearance of the elements of the G type culture, in passages of the S or the R cyclostages through plain or modified broth, detailed data have already been presented on an earlier page.[30] But exactly the same circumstances appear in relation to the first observed appearance of the bacteriophage in a series of tubes undergoing repeated feedings (with sensitive culture) and filtrations. In some still unpublished experiments on the "spontaneous" appearance of the bacteriophage, performed by two of us (J. K. and P. H.) and our associate, Edna Kiesewetter, some years ago, the following results appeared: In attempts to generate bacteriophage for the Shiga bacillus by repeated growth of the stock S type culture in its own filtrates and by the use of the same culture and same methods in all tests the positive result was obtained once after the twentieth serial filtration and twice after the fourteenth. Two of us (P. H. and J. K.), moreover, found that when pancreatin broth was used as a medium and the same method of alternate feeding and filtration employed, the bacteriophage was spontaneously generated with remarkable regularity in about the eighth serial filtrate. In the present cases it has appeared that the G forms were generated most commonly between the sixth and the twelfth serial tube—in several instances, in the tenth—both in the Shiga bacillus and in *B. typhosus*. We are therefore led to attach considerable significance to the definite "Momenten" which, under constant conditions of experiment, both define and limit the time of spontaneous appearance, not only of the elements of the G type cultures, but also of the corpuscles of the bacteriophage. We must also bear in mind this

---

30. For evidence of this fact, the reader may consult the origin of the G type cultures, strains I and Ia, II and IIa, in tables 1 and 2 of the present report.

highly important fact: The elements immediately preceding the appearance of the G forms are apparently noncultivable; and they are often associated, in point of time, with the disappearance (apparent destruction) of the S or the R type cells previously present in numbers. It is thus suggested that the bacteriophagic corpuscles are in some still unrecognized manner associated with the filtrable elements that are the predecessors of the G forms. In any case, we must regard both the so-called "lytic units" and the G elements as present potentially in the "normal" culture; and recognize, further, that the same influence (enforced dissociative reaction) serves to liberate both. What is the actual relation existing between the lytic corpuscles and the filtrable elements preceding the G type culture? This question, it appears to us, introduces the crucial point of investigations of the nature of the bacteriophage.

Continuing our survey of the similarities between the bacteriophage and the filtrable elements associated with the G type cultures, it may be added that both seem to be equally filtrable through Berkefeld candles. Moreover, while the corpuscles of the bacteriophage are commonly believed to be noncultivable in or on any medium, the filtrable bodies of the G type culture (and especially those passing the N and W candles) are not, as a rule, visibly cultivable on the first agar plate seeded with the filtrate and often not even on the second or third. Broth filtrates containing the virus forms in abundance remain water-clear for weeks or months. Little by little, however, these elements become susceptible of cultivation in a visible form and, eventually, they regain the original G type. In a similar manner we have demonstrated, in experiments not recorded in this paper, that the majority of our lytic filtrates, active on Shiga and typhoid strains, sealed in glass ampules and at first apparently sterile, seldom fail, after storage of weeks or months, to yield cultures that appear identical with the G type cultures of the homologous species.

What is the relation of the bacteriophage to these filtrable, but eventually cultivable, organisms that arise, in time, in the lytic filtrates? What is its relation to the filtrable forms of the G type cultures, obtained without the recognized intervention of bacteriophage acting on the original cultures? What is the relation of the filtrable and cultivable bodies obtained from lytic filtrates to the filtrable and cultivable elements obtained from cultures experiencing the dissociative reaction? When these questions are answered, the nature and origin of the bacteriophage will be less of a mystery.

RELATION OF G FORMS TO METHODS OF RECOGNIZING STERILITY OF
CULTURES AND OF PATHOLOGIC TISSUES

One of the important bearings of this study in the field of applied bacteriology concerns criteria for, and methods of detecting, the sterility of cultures, culture mediums and normal or pathologic tissues. Our older teachings in bacteriology have been such as to imply that, although there may exist "noncultivable" organisms and organisms that produce an "invisible growth," the commonly accepted indications of sterility of a medium are failure of clouding or of sedimentation in liquid mediums and absence of visible foci of growth on solid mediums after a suitable period of incubation. It is usually brief observations of this sort that have been made when the writer states that certain mediums were "tested for sterility," or that cultures "proved sterile," or that cultures "failed to yield growth" on transplantaion.

From the observations made during our study of the G type cultures of the Shiga bacillus, we have reached the opinion that the demonstration of the actual sterility of a substance or medium or tissue is a matter demanding far more careful attention and a vastly more perfect technic than have been accorded this problem in the past. Failure of apparent growth in liquid or on solid mediums may indicate nothing so far as actual sterility is concerned, except, perhaps, that the particular form of growth that one has been accustomed to search for is absent. It does not necessarily mean that living elements of a sort, representative of some bacterial species, and capable of yielding, in due time and under favorable conditions, the specific features of the microbe, are absent.

This circumstance, it may be imagined, has an important bearing in many fields of bacteriologic practice, of which we will here mention only one. This concerns the differentiation between the action of certain bacterial toxins and the action of filtrable forms of the organism in question. This matter has been introduced by Ramsin (1926), dealing with the scarlet fever toxin. This, he alleged, has been confused with filtrable forms of the scarlet fever streptococcus. It also has a bearing on the production and employment of the so-called "in-vivo-prepared toxins," such as those reported for *B. typhosus* and *B. coli* by Harris. From unpublished experiments which one of us (P. H.) has performed with his students, Frohman and Ratner, with *B. typhosus,* and from others performed by ourselves with the Shiga bacillus, we are inclined to doubt that it is possible to obtain sterile Berkefeld filtrates as a result of filtering the peritoneal exudates obtained from laboratory animals previously inoculated with broth cultures or salt solution suspensions of the typhoid bacillus, the Shiga bacillus and related organisms. It is true that the plating out of such filtrates on agar, the cultivation of even larger amounts in broth, or even incubating the original filtrates themselves as a test for sterility is likely to yield none of the common indications of the presence of living bacteria. Conclusions as to the sterility of such filtrates would seem to be justified, but in reality they may be open to grave error. A special technic for the demonstration of sterility must be employed in these cases; and, it should be added, even if living organisms are discovered, they may not appear to be related, so far as colonial, cultural or morphologic characters are concerned, to the original strain employed in the investigation. It is by no means certain that some of the striking effects following the injection of the "in-vivo-prepared toxins" and other toxic filtrates into animals are not due to the presence and multiplication, in the bodies of the inoculated animals, of a filtrable stage in the life history of the organism concerned.

We do not wish to be misunderstood in implying that we believe there is an identity between the scarlet fever toxin and the filtrable form of the streptococcus, between the typhoid toxin and the filtrable form of the typhoid bacillus (an entity which has certainly been proved to exist) or between any other bacterial toxin and a filtrable form of the homologous culture. Our intention in making these references is merely to call attention to the circumstance that, in certain cases, it may be a more difficult task than commonly believed to differentiate clearly, by purely biologic methods, between these two products of bacterial growth, namely, toxins and filtrable stages in the life history of the organisms concerned.

POSSIBLE RELATION OF FILTRABLE FORMS OF BACTERIA TO ETIOLOGY OF COMMUNICABLE DISEASES

If, as we believe we have demonstrated for the Shiga bacillus, there exists, at a certain moment in the life of the culture, a minute, probably invisible and filtrable form of the organism, yet one capable of developing into a form that is able to propagate itself on common culture mediums, although not yielding, for a time at least, the usual evidences of cultivability; and if, as we or other members of our laboratory group have been able to demonstrate in B. coli, B. typhosus, B. paratyphosus A, B. paratyphosus B, B. enteritidis, B. cholerae-suis, B. typhi-murium I

B. typhi-murium II (*B. pestis-caviae*), the diphtheria bacillus (Park 8), the cholera vibrio, and in *B. acidophilus,* such filtrable bodies can be isolated from common laboratory cultures or, in some cases, from the blood and tissues of animals infected with filtrates of G. type cultures, it is of interest to consider what rôle these filtrable bodies might play in active disease; also the extent to which analogous living elements might be anticipated in the life history of still other, or perhaps all, bacterial species.

From a wide range of observation during the last two decades, and from more detailed studies of the past eight years, the conclusion has been forced on the few bacteriologists who have given serious attention to these matters, that the relation of a given bacterial species to a specific disease involves certain complex aspects, previously accorded little attention either by the laboratory worker, clinician or epidemiologist. For exactly fifty years, our notions regarding these matters have progressed but an imperceptible distance beyond the stage first depicted by Koch when he wrote in his classic treatise on wound-infection, in 1878. "A special form of germ is associated with every disease, and this form remains unchanged irrespective of the number of times the infection may be transferred from one animal to another."

Exceptions to this general rule have, in more recent years, appeared in the works of a small number of writers, including Hort, Mellon, Zlatogoroff and Neufeld. All of these investigators, refusing to be bound by old traditions, have performed experiments through which they have clearly recognized the important rôle that variability within the species may play in the course of infection, in epidemiology and in pathology. It is impossible, within the confines of the present paper, to review the observations and conclusions of these workers or the important work of Löhnis, which, though less in the field of pathogenic bacteriology, supplies evidence of a similar nature. A brief résumé of some of the more important points of their studies may, however, be given; for their experiments and conclusions are inextricably bound up with the dissociation problem, and therefore, in a measure, with that of the filtrable forms of bacteria.

Hort's work in England, in the years 1914 to 1916, concerned chiefly the meningococcus and the typhoid bacillus (1916, a and b), but the principles on which he based his observations were more general (1916, c). He was unable to produce cerebrospinal fever in monkeys by the injection of cultures of the meningococcus. It is now commonly admitted that Flexner's inoculations of culture into these animals likewise produced only meningitis, and that his results were due mainly to the toxic action accompanying the large doses employed. But Hort was able to show, further, that filtrates of the spinal fluid of patients contained a virus which had the power of causing a continuous fever in monkeys, and which gave rise, in the bodies of the infected animals, to the typical organism of Weichselbaum, as well as to the Jaeger coccus. These findings led Hort to the view that the true virus of cerebrospinal fever is a filtrable stage in the life history of the meningococcus. These results and conclusions, first presented more than fifteen years ago, have never been confirmed or denied, and the meningococcus still retains its traditional position. To the aforementioned observations should be added the highly suggestive work which Hort accomplished on typhoid, even going so far as to indicate the possible mode of origin of certain minute, and probably filtrable, forms of the bacillus. Hort's labors in these and related fields, brought to an untimely end, have served to place him in the front rank of investigators devoted to these problems.

Mellon, in 1920, in his paper, "Life Cycles of Bacteria and Their Possible Relation to Pathology," gave voice to certain views which, though at that date sufficiently timely, have gained further support through investigations performed both by himself and others in later times. In the study of a certain diphtheroid organism, Mellon noted the passage of the culture through various phases of growth, including diplococci, giant cocci, streptococci, long, filamentous forms, bacillary forms and filtrable bodies. In a culture of *Streptothrix* he showed that a filtrable form, present in the blood of the patient, could be cultivated on artificial mediums as a diplococcus, which, in turn, could be transformed into the filamentous or branching form. A giant coccoid stage was also present, as in the diphtheroids. A second form of filtrable body was found for the diphtheroid stage of the *Streptothrix*. Mellon regarded these elements as comparable with the Much granules of the tubercle bacillus. They were definitely virulent, while the bacillary forms were innocuous. Mellon also reported that the branching phase gave rise, by budding, to filtrable bodies which could not be cultivated beyond the first generation. Still later, Mellon described a virulent and filtrable stage of the fusiform bacillus and believed that it represented the gonidial form.

Mellon pointed out the possible significance of these and other similar observations for our views on the etiology of hog cholera, Hodgkin's disease, scarlet fever, typhus fever, poliomyelitis and influenza. He recalled that Noguchi found small coccoid bodies in his cultures of *pallida,* and that Warthin observed similar elements in tissues devoid of spirochetes, but unquestionably syphilitic. Some of these we ourselves have observed in Warthin's preparations.

Zlatogoroff, in his brief but important contribution published in 1924, also recognized the extensive range of bacterial variation and debated whether the variants should be regarded as mutants in the strict sense of de Vries; also whether the modified forms should be considered as independent bacterial species or as varieties of species already known, or, again, whether the variants should be regarded as representing an evolution of forms in which the character-limits of the single group were being transcended. To answer these questions he advocated particularly a study of the origin of the variations. He drew attention to the far-reaching modifications that may arise in the germs derived from the body of the infected animal, as well as under normal conditions, and cited the work of his laboratory on the cultural, serologic and bio-chemical variations in the cholera vibrio. He pointed out that all the varieties observed must preexist in the protoplasm of the mother cul-ture, and, conversely, how the mother culture must be preexistent in the several variants of each species. Unfortunately, Zlatogoroff was inclined to explain these variants as concerning either the laws of Mendel or mutation phenomena, both of which he believed he could see operating in the world of the bacteria, as well as in higher forms. One of us (P. H.) has already had occasion to note in an earlier paper (1927) the entangle-ment of ideas that usually results when bacteriologists undertake to explain variation phenomena in bacteria in terms of currently held genetic principles. There has always been a strong tendency to grasp at any floating theories, about which even the geneticists themselves are none too certain, rather than to study what is actually going on in the genetic activity of bacterial forms. It is always easier to call a thing a "mutation" than to investigate its nature and origin otherwise. The explanations of these phenomena offered by Zlatogoroff do not, however, detract from the importance of his observations; they have merely limited the range of visibility. He stood for the fact that vari-ation phenomena of a sort, among bacterial species, whatever the cause may be, underlie some of the most important problems in bacteriology and pathology, and this alone marks a far advance from current bac-teriologic conceptions.

Neufeld, in his de Lamar lectures, in 1926, on variability in bacteria, again struck at the root of the problem, though without recognizing the place of his variants in the cyclogeny of the species or their relation to the phenomenon of dissociation. His observations, which concern mainly the variability of micro-organisms occurring in the body of the host, led him to modify the older notions of Koch (1878) relating to the specific etiology of infectious disease. He well pointed out that a doctrine of such fundamental importance naturally cannot help, in the course of time, being taken in too rigid and dogmatic a sense, and he was inclined to modify the older view that the form of infecting organism remains unaltered no matter how many times the disease is transmitted. Regarding the nature of the variations themselves, Neufeld did not relate them to what we now term dissociative action, but regarded them as evidence of a degeneration of the organisms concerned. This, he believed, is the case with culture changes observed in *B. diphtheriae,* streptococci and pneumococci, when they become transformed from virulent or toxigenic into nonvirulent or nontoxigenic forms.

It is also of interest that Neufeld accepted the existence of filtrable forms. He stated (de Lamar Lectures, p. 18), "In the course of such variation processes bacteria may produce minute, filter-passing modifications." He mentioned the unpublished work of Levinthal in which it was shown that, in the case of *Bact. pneumosintes,* it was possible to change the microbes from a minute, filtrable form to a form resembling the Pfeiffer bacillus merely by modifying the nutrient medium. Neufeld believed, as the result of his own studies and those of his collaborators, that bacteria undergo highly significant changes from the moment when they invade the body of the host; that they may not only "degenerate," but also acquire new properties, although he admitted that, in the last instance, the acquisition of the "new properties" may mean merely that the culture regains an old characteristic that "may have remained in them potentially in a latent form." Although Neufeld believed that the peculiar transitions could be explained only on the basis of some degenerative change, and although he did not recognize in these transitions strictly "normal" dissociative reactions, no other writer, aside from Mellon, Zlatogoroff and Enderlein, has approached so close to the heart of the variation problem as it bears on the relation of the various cyclostages of pathogenic bacterial species to the disease process in man and animals. All this is of special interest when it is borne in mind that these "radical" views have emanated from the ancient birthplace of monomorphism.

Such views as those voiced by Mellon, Hort, Zlatogoroff, Neufeld, Löhnis, Almquist, Enderlein and a few others have never been popular among bacteriologists, especially those in America. But it must be admitted that new observations are constantly being added to support the conclusion that the conditions present in the body surrounding bacterial infections are conditions that instigate bacterial variability, and that the common bacterial forms isolated from cases of infectious disease may not always be the exact agents instrumental in producing or in maintaining the infection. In other words, the microbes that are easily isolated and cultivated from the blood or tissues may not always be secondary cultures from the point of view of the specific infection, but they may be secondary from the point of view of the cyclogeny of the species. This conception cannot readily be proved, although it does not lack supporting evidence. To say that the species *B. typhosus* causes typhoid fever, that the species *Meningococcus* causes cerebrospinal fever or that a species of hemolytic streptococcus causes scarlet fever is probably correct. To say, on the other hand, that all the clinical symptoms and pathology of one of these maladies are determined directly and exclusively by that form of the organism usually obtained in laboratory cultures and designated and pictured in the textbooks as the causative agent of the disease in question is to make a statement that is likely to be very far from the truth. In other words, there is being furnished in recent literature ever increasing evidence for the view that a "bacterial species" can no longer be regarded as made up of homologous units, each resembling the other in form, cultural features and physiologic action. While stability may be present to an extent and for a brief time in laboratory cultures—in which we are dealing with domesticated cells—the species in its entirety must be regarded as comprising a far greater range of variation than is commonly recognized. In a single tube and at a given time, the culture mass may seem to constitute a homogeneous population; but, if given the freedom for continued development (as, for example, in serial cultivations or in large volumes of medium) and if carefully observed, it appears to us as a highly heterogeneous population of living cells which manifest, over a period of time, significant and progressive differences in form and physiologic capacity. Whether we have in our hands a pure culture of the S type or of the R type Shiga, it is the Shiga dysentery species; and if, instead of these, we have the filtrable G type, it is still, potentially, the Shiga species. To regard this microbe as comprising only the classic, textbook form (invariably the S cyclostage) gives us almost as inadequate a notion of the complete species as if we limited our recognition of *Plasmodium malariae* to the stage of development observed in the red blood cells of the human host or to the elements present in the salivary glands of the mosquito.

We are, moreover, beginning to see more clearly that these variations in bacteria possess a significance quite different from the "fluctuating variations" in the Darwinian sense, from the "mutations" of de Vries and from the "youth to age" variations described in recent literature. They possess, moreover, no characteristics that would entitle them to the appellation "hereditary variations" in the sense of Arkwright, since they are not, in the strict sense, hereditary.[31] Their actual significance lies in the circumstance that each variation has its own definite position in the ontogeny of the species. The transformation of one form into another does not appear to be a chance occurrence; nor one determined wholly by environmental conditions, although these conditions may hasten or retard the directive trend. The development follows an orderly course, and this course is often pursued, not because of, but in spite of, the nature of the environment. Moreover, these transformations may require time—not always hours, but often days, weeks or even months. While such transformations usually occur naturally and slowly, they may be forced to appear more rapidly by the application of adequate stimuli, either inside the body of the host or in the tube. But, whether occurring naturally or forced by definite reagents or other growth conditions, the broader trends and the outcome of the process are apparently the same in both cases.

Now the important bearing of these matters on the present aspect of the subject is that the ability to produce disease is not possessed equally by all stages in the cyclogeny of any pathogenic species. In most cases that have been studied in this regard it appears to be the S form that harbors the greater virulence, as also the greater toxicity. In *B. anthracis,* on the other hand, and probably in some of the streptococci (both hemolytic and greening), there is evidence that the R type is as virulent as, or perhaps more so than, the S. In streptococci from root-canals of infected teeth the R form is more virulent than the S, as recently shown by Faith Hadley. In *B. fusiformis* and in certain species of *Streptothrix,* as observed by Mellon (1926), it is suggested that a

---

31. We are unable, in this connection, to subscribe to the views recently presented by Arkwright on the "hereditary variations" of bacteria as opposed to the concept of cyclostages. This author's presentation escapes the use of the term "mutation," as certainly appears desirable, but we do not see that the term "hereditary variation," as applied to the bacteria, carries an essentially different meaning.

filtrable stage harbors more invasiveness than do the vegetative rods and threads. In other words, it has become fairly clear that only a single cyclostage harbors the maximum virulence or toxicity. The other cyclostages carry merely what might be termed "potential virulence."

The question therefore arises: If the virulence of microbes of different pathogenic species attaches sometimes to one cyclostage, sometimes to another, depending on the specific cyclogeny of the organism concerned, what may be the pathogenic significance of those filtrable forms of culture which have been produced and artificially cultivated in the case of the Shiga bacillus and several other forms? Also, what may be the pathogenic significance of the filtrable forms of other bacterial species which, we believe, will soon be brought to light? Further, do there exist bacterial species whose commonly recognized forms are harmless, but whose still unrecognized, filtrable forms are pathogenic? Aside from a few cases possessing considerable suggestiveness, these questions must remain for the present unanswered. Our own experiments dealing with the filtrable forms of the Shiga bacillus are still in progress. At the present moment it is indicated that the G. culture type is lacking in toxicity and virulence and may possess immunizing power against the toxic S cyclostage. The new field of study with this and other species is alluring, and it seems to us, may not be destitute of significant issues in relation to certain disease of still unknown etiology.

POSSIBLE RELATION OF FILTRABLE FORMS OF BACTERIA TO STUDY OF FILTRABLE VIRUSES

The present trend in the study of the filtrable viruses is apparently to regard them as comprising an independent order of living beings, autonomous, usually parasitic and distinct from known bacterial species; in other words, to place them in a class by themselves, as d'Herelle (1922) did, under the group "Protobios," the most primitive of living things. With this conception in mind, the ultraviruses are pictured as autonomous, but having a nature and organization much simpler than that of the known bacteria. Indeed, to d'Herelle, such minute particles of living protoplasm are not to be regarded as true cells in the usual sense, but as protein micellae, possessing complete autonomy.

On the other hand, it has been tentatively suggested—most clearly, though somewhat apologetically, by Nicolle (1925)—that the filtrable viruses may be only the invisible stages of microscopically visible microbes, into whose forms and attributes the viruses may, under especially favorable conditions, become retransformed. Which of these two hypotheses is likely to harbor the larger element of truth will probably not be known for many years. The results reported in this paper favor one possibility no more than the other. As a matter of fact, it requires no great stretch of imagination to suspect that both might be true; that there exist bacterial diseases in which the actual causative agents are virulent, filtrable cyclostages, possessing a transient stabilization; also that there exist other communicable diseases in which the causative agents are minute corpuscles, analogous perhaps to the filtrable forms of bacteria, but differing from them mainly in respect to the point that the gonidia-like, virus particles have, in the course of their evolution, become sufficiently stabilized to maintain their continued autonomous existence, independent of the higher and more commonly observed cyclostages of the bacterial culture. Looking at the matter purely from the point of view of bacterial phylogeny, one might surmise (according to one's biologic inclinations) either that the virus particle represents a degenerate, but at the same time free-living and independent, form of a higher bacterial type; or that it represents the primitive, gonidium-like unit from which through long, evolutionary processes the higher and microscopically visible forms of the bacteria have arisen. From a purely biologic standpoint, it does not appear that one conception has more to recommend it than the other, since evolutionary changes may be either progressive or regressive.

If, however, one assumes that the higher fungi have evolved from simpler and independent cell forms, there is no reason to believe (aside from the limitations that the modern microscope enforces on one) that these primitive, bacteria-like organisms themselves have not evolved from forms still smaller and still more primitive; in other words, that, in the obscure evolutionary background of the bacteria (which one is pleased to call the smallest and simplest of known plant cells) there have existed, and perhaps still exist, minute beings from which the present recognized races of bacteria, in turn, have evolved, and which might be characterized as minute, free-living, gonidia-like forms. If such a primitive, living element exists, or has existed, all the established principles of phylogeny serve as the basis of a prediction that this organism should be represented by a form analogous to the most primitive of the cyclostages now recognized in any bacterial species; and this form, without possible question, is the "seed form" or bacterial microgonidium.

On this interesting and important problem regarding the possible relationship of the filtrable forms of bacteria to the large group of filtrable viruses, we believe that our observations can throw little light. The results that we and our associates have obtained reveal merely the fact that several species of common pathogenic bacteria are able to enter a stage of development in which they are easily and invariably filtrable, presumably ultramicroscopic and, at least in part, and for a limited time, noncultivable (visibly) in or on common laboratory culture mediums; moreover, that success in transforming the virus stage into the common visible form demands a special technic rather different from that commonly applied.[32] This in itself does not appear to be much of an advance in knowledge, but it perhaps carries a significant implication. Even if this much is true, and if it could be demonstrated that the facts involved possess a general application to all bacterial species, it is possible that we should prepare ourselves to adopt a fundamentally different attitude toward the problem of the nature and mechanics of certain little-understood microbic infections. Perhaps

---

32. For example, it would not be anticipated that the methods of cultivation employed by certain workers in their attempt to generate visible forms of strep-tococci of the Evans encephalitis type from filtrates of brain material of virus-infected rabbits would be successful. Negative results obtained by the use of the commonly employed methods can lead to no definite conclusions one way or the other.

In this connection it may be added that in 1929 one of us (P. H.), in collaboration with Mr. Weaver of this laboratory, succeeded, by using the methods delineated in this paper, in isolating from glycerolated samples of the Levaditi and the H. F. herpes-encephalitis virus a greening streptococcus. The same culture was obtained from Berkefeld N filtrates of the latter virus. The filtrable form, in the course of reversion, passed through a stage in which it failed to produce greening on blood agar plates and grew better at room temperature than at 37 C. At a certain early stage in reversion, the cultures from the filtrates of the H. F. virus were more infectious for rabbits (from 0.2 to 0.3 cc. killing in from nine and one-half to sixteen hours) than the original virus (0.3 cc. killing in five and one-half days).

Regardless of the possible relation of the greening streptococcus to herpes and encephalitis, it appears that a certain stage in the cyclogeny of a greening streptococcus is commonly present in rabbit and guinea-pig brain viruses, and that in certain cases the organism is present in a filtrable form. It therefore seems to us reasonable to doubt that a clear case can be made out in favor of the existence of an encephalitic virus of independent biologic nature until the filtrable forms of certain greening streptococci have been diligently searched for by approved methods, and their possible lethal rôles eliminated from the field of argument.

we should be led to envision the probability that living germs are present and perhaps active in organs and tissues that reveal few or none of the ordinary indications of bacterial infection; also the possibility that germ life of a sort may be the causal factor in certain communicable diseases in which, as yet, we have been able to gain no definite knowledge of the presence of bacterial infections in the usual sense or by the customary methods.

Indeed, so long as investigators continue, voluntarily or involuntarily, to close their eyes to the existence and vast significance of dissociative reactions among bacteria at large and arbitrarily (without the willingness to perform, carefully and intelligently, the crucial experiments) deny the existence of virus-like forms embraced within the complete cyclogeny of both pathogenic and nonpathogenic bacterial species, they may be failing to utilize the most promising avenue of approach, not only to a deeper understanding of the biology of the bacteria, but also to the study—and perhaps solution—of a multitude of important but still obscure problems in the fields of bacteriology, pathology and medicine.

GENERAL CONCLUSIONS AND SUMMARY

In this investigation we have, for the first time we believe, generated artificially, and subsequently cultivated in pure lines, the filtrable, virus-like stage of a bacterial species. We produced this form (G type culture) by forcing the dissociative reaction on the mother S or R type culture. We grew this G form in pure cultures as a distinct cyclostage before demonstrating that at least some of the cellular elements contained in it were invariably filtrable through Berkefeld N and W candles, as opposed to the S and R type cells, which are not filtrable under the same conditions.

From the filtrates we were successful in recovering the G type culture. We have demonstrated, moreover, that the visible elements of the filtrable phase are very different culturally, morphologically, biochemically, serologically, immunologically and in their relation to the Shiga bacteriophage from the "normal" forms of the Shiga species, but that they may be caused to "revert" to the original cell type, possessing all of the original characteristics, including toxicity and susceptibility to bacteriophagic influence.

We have shown that the filtrable bodies experience a phase of existence in which they are not visibly cultivable in or on the usual culture mediums and by the usual methods, and that a special technic, such as that used by Hauduroy, must be employed for bringing them into visible development. We have demonstrated, in addition, the high degree of stability enjoyed, in many cases, both by the G type cultures and by the associated virus-like forms when sealed in ampules and stored for periods exceeding two years; we have come to regard them as the most stable forms of the Shiga species.

Besides demonstrating the filtrable stage for the Shiga bacillus, we or other members of our laboratory group have shown the existence of analogous phases of culture development in the ontogeny of ten other bacterial species. Their G forms are equally filtrable.

Regarding the nature of the G type cultures and the associated filtrable forms, we have concluded that they represent definite cyclostages in the ontogeny of the Shiga species, and that they constitute, in part at least, the gonidia and the microgonidia. We regard the G type cultures as the visible stage of culture existence lying between the noncultivable (visibly) virus form and the ordinary culture type (S). Both, therefore, have a definite relation to microbic dissociation and are natural products of the dissociative mechanism.

The exact cytologic origin of the G forms and the associated virus bodies is not yet entirely clear. There may be several modes of origin, and these may not be the same in different bacterial species. The only origins of which we can speak with any degree of assurance are the lateral or terminal "buds" formed on the rods of old S cultures and the granular inclusions ("gemmules," "gonidia") produced in, and liberated by, the long rods and the filamentous structures that characterize the R type culture. Among the species that we have studied we regard this fungoid R phase as the reproductively mature culture form. The S, we believe, represents a lower, vegetative stage in the ontogeny of the species, while the G type, with its associated virus-like bodies ("microgonidia"), represents the lowest. The specific function of the S type cells is, therefore, vegetative growth; that of the R type cells, reproduction; that of the G forms, dissemination of the species.

Regarding the significance of these new filtrable forms for bacteriology, epidemiology and pathology, although much must be left for future study, we have felt justified in discussing in a tentative manner some of the possibilities that have occurred to us. At the same time we observe that the filtrable phase of ontogeny is not limited to the intestinal bacteria; and it is possible that the analogous forms of other species may possess a different significance from that suggested by the Shiga bacillus, particularly from the point of view of virulence.

Regarding the possible relation of these virus-like bodies to the large group of filtrable viruses, we believe that our results offer no grounds for conclusions. At the same time, they perhaps suggest that it will be of advantage, in the future, to incorporate the point of view of microbic dissociation in the study of certain so-called virus diseases.

BIBLIOGRAPHY

Albrecht, R.: Beitrag zur Kenntnis der Entwicklung der Spirochaete obermeiri, Deutsches Arch. f. klin. Med. **29**:77, 1881.

Almquist, E.: Zur Biologie der Typhusbakterie und der Escherischschen Bakterie, Ztschr. f. Hyg. u. Infektionskrankh. **15**:283, 1893.
Studien über filtrierbare Formen in Typhuskulturen, Centralbl. f. Bakteriol., I, O. **60**:167, 1911.
Wuchsförmen, Fructification und Variation der Typhusbakterie, Ztschr. f. Hyg. u. Infektionskr. **83**:1, 1916.

Arkwright, J. A.: Section on Variation in System of Bacteria, London, H. M. Stationary Office, 1930, pp. 320-321.

Arloing, F., and Dufourt, A.: Récherches sur le pouvoir infectant des filtrats de tuberculose avaire employés en injections chez le pigeon et la poule, Compt. rend. Soc. de biol. **101**:455, 1929.

Babes, V.: Comparaison entre les bacilles de la tuberculose et ceux de la lèpre, Compt. rend. Acad. d. sc. **96**:1323, 1883.

Balfour, A.: The Infective Granule in Certain Protozoal Infections, Brit. M. J. **1**:752, 1911.

Beijerinck, M. W.: Die Bakterien der Papilionaceenknollchen, Botan. Ztschr. **46**: 725, 1888; cited after Löhnis, F.: Studies on the Life Cycles of the Bacteria, Mem. Nat. acad. d. sc. **16**:1, 1921.

Billroth, T.: Untersuchugen über die Vegetationsformen der Coccobacteria septica, Berlin, 1874.

Bordet, J.: La morphologie du microbe de péripneumonie des bovidés, Ann. de l'Inst. Pasteur **24**:161, 1910.

Borrel; Dujardin-Beaumetz; Jeantet, and Jouan: Le microbe de la péripneumonie, Ann. de l'Inst. Pasteur **24**:168, 1910.

Breinl, A.: On the Morphology and Life History of Spirochaeta Duttoni, Ann. Trop. Med. **1**:433, 1907.

——and Kinghorn, A.: An Experimental Study of the Parasite of African Tick Fever (Spirochaeta Duttoni), Mem. Liverpool School Trop. Med., 1906, no. 21, p. 1.

Bronfenbrenner, J., and Muckenfuss, R.: Proc. Soc. Exper. Biol. & Med. **44**:371, 1927.

Burdon-Sanderson: The Origin and Distribution of Micro-Organisms in Water, etc., Quart. J. Micr. Sc. **11**:323, 1871; cited from Löhnis (1921).

Burnet, E.: Actions d'entrainement entre races et espéces microbienne, Arch. Inst. Pasteur de Tunis **14**:384, 1925.

Sur la récherche de formes filtrantes des bactéries, Compt. rend. Soc. de biol. **95**:1142, 1926.

Streptocoque de sortie chez le lapin, ibid. **98**:440, 1928.

Calmette, A., and Valtis, J.: Ann. de méd. **19**:553, 1926.

Cienowski, L.: Zur Morphologie der Bakterien, Mem. Acad. St. Petersburg, 1877, vol. 25, no. 2; cited from Löhnis (1921).

Cohn, F.: Beiträge zur Biologie der Pflanzen, Breslau, 1875, vol. 1, p. 127.

Cooper, F. B., and Petroff, S. A.: "Filtrable Forms" of the Tubercle Bacillus, J. Infect Dis. **43**:200, 1928.

Cunningham, D. D.: Choleraic and Other Commas, etc., Scient. Mem. Med. Officers, Army of India, 1897, vol. 10, p. 1; cited from Löhnis (1921).

Dowdeswell, G.: Sur quelques phases du dévelopment du microbe du choléra, Ann. de micrograph. **2**:529, 1889.

Dutton, J. E., and Todd, J. S.: The Nature of Human Tick Fever, Mem. Liverpool School Trop. Med., 1905, no. 17, p. 1.

A Note on the Morphology of Spirochaeta Duttoni, Lancet **2**:1523, 1907.

Ehrenberg, C. G.: Die Infusionstierchen als volkommende Organismen, Leipzig, 1838.

Eisenberg, P.: Ueber die Anpassung der Bakterien an die Abwehrkraft des infizierten Organismus, Centralbl. f. Bakteriol., I, O. **34**:739, 1903.

Enderlein, G.: Bakterien-Cyclogenie. Prolegomena zu Untersuchungen über Bau, geschlechtliche und ungeschlechtliche Fortpflanzung und Entwicklung der Bakterien, Berlin, W. de Gruyter & Company, 1925.

Evans, Alice: Studies on the Etiology of Epidemic Encephalitis: I. The Streptococcus, Pub. Health Rep. **41**:1095, 1926.

Ewart, J. C.: (a) The Life History of Bacterium Termo and Micrococcus, Proc. Roy. Soc., London **27**:474, 1878.

(b) On the Life History of B. Anthracis, Quart. J. Micr. Sc. **18**:161, 1878.

Fantham, H. B.: Some Researches on the Life Cycles of Spirochaetes, Ann. Trop. Med. **5**:479, 1911.

Fejgin, B.: Sur les variations brusqués du Proteus HX19 survenue sous l'influence de l'agent lytique anti-HX19, Compt. rend. Soc. de biol. **90**:1106, 1924.

Sur la forme filtrant de bacille d'Eberth, ibid. **92**:1528, 1925.

Fokker, A. P.: Milzbrand ohne Stäbchen, Centralbl. f. d. med. Wissensch. **9**:20, 1881.

Fontès. A.: Estudos sobre a tuberculose, Mem. do Inst. Oswaldo Cruz, **2**:186, 1910.

Friedberger, E., and Meissner, G.: Sur Pathogenese der experimentelle Typhus Infektion der Meerschweinchen, Klin. Wchnschr. **2**:450, 1923.

Frobisher, M.: On the Action of the Bacteriophage in Producing Filtrable Forms and Mutations of Bacteria, J. Infect. Dis. **42**:461, 1928.

Frobisher, M., Jr., and Brown, J. H.: The Toxicogenic Properties of Some Streptococci, J. Bact. **13**:44, 1927; Bull. Johns Hopkins Hosp. **41**:167, 1927.

Fuhrmann, F.: Entwicklungscyklen von Bakterien. Verhandl. d. Gesellsch. deutsche Naturf. u. Arzte **78**:278, 1906.

Garrod, L. P.: Filter Passing Anaerobes in the Upper Respiratory Tract, Brit. J. Exper. Path. **9**:155, 1928.

Geddes, P., and Ewart, J. C.: On the Life History of Spirillum, Proc. Roy. Soc., London **27**:481, 1878.

Gratia, A.: Héterogéneité du principe lytique du colibacille, Compt. rend. Soc. de biol. **89**:821, 1923.

——and de Kruif, Lois: Tentative d'isolément de bactériophage d'inégale activité, Compt. rend. Soc. de biol. **88**:629, 1923.

Gunther, C.: Einfuhrung in das Studium der Bakteriologie, Leipzig, 1906.

Haag, F. E.: Der Milzbrandbacillus, seine Kreislaufformen und Varietäten, Arch. f. Hyg. **98**:271, 1927.

Hadley, Faith P.: The Relation of Virulence to Colony Variation in the Streptococci, J. Am. Dent. A. **17**:1730, 1930.

  Recognition of Bacillus Acidophilus Associated with Dental Caries, J. Am. Dent. A. **17**:2041, 1930.

Hadley, Philip: Parallelism Between Serologic Response and Bacteriophagic Response in B. Typhosus and Certain Avian Paratyphoids, Proc. Soc. Exper. Biol. & Med. **23**:443, 1926.

  Microbic Dissociation, J. Infect. Dis. **40**:1, 1927.

  The Twort-d'Herelle Phenomenon, ibid. **42**:263, 1928.

——and Dabney, Eugenia: The Dual Nature of the Lytic Principle: a Study of the Alpha and Beta Units of a Paratyphoid Bacteriophage, Proc. Soc. Exper. Biol. & Med. **25**:355, 1928.

——and Klimek, John: The So-Called "Origin" or "Source" of the Bacteriophage, Proc. Soc. Exper. Biol. & Med. **25**:34, 1927.

——and Jiménez, Buenaventura: Production of Bacteriophage by Enforced Dissociation, to be published in J. Infect. Dis., Feb., 1931.

Harris, W. H.: Experimental Typhoid Fever Induced in Guinea-Pigs with In-Vivo-Prepared B. Typhosus Toxic Products, Proc. Soc. Exper. Biol. & Med. **25**:372, 1928.

Hauduroy, Paul: (a) Présence des formes filtrantes du bacille d'Eberth dans le sang d'une typhique, Compt. rend. Soc. de biol. **95**:288, 1926.

  (b) Les formes filtrantes des bactéries et les ultravirus, ibid. **95**:1523, 1926.

  (a) Techniques de culture des formes filtrantes invisibles des microbes visibles, ibid. **97**:1392, 1927.

  (b) Toxines diphtériques donnant naissance à un bacille diphtérimorphe, Compt. rend. Acad. d. sc. **184**:409, 1927.

  Les ultravirus et les formes filtrantes des microbes, Paris, Masson & Cie, 1929, pp. 1-392.

——and Lesbre: Les formes filtrantes des streptocoque, Compt. rend. Soc. de biol. **97**:1394, 1927.

Henrici, A. T.: Morphologic Variation and the Rate of Growth of Bacteria, Springfield, Ill., Charles C. Thomas, 1929.

d'Herelle, F.: The Bacteriophage and Its Rôle in Immunity, tr. by George H. Smith, Baltimore, Williams & Wilkins Company, 1922.

  The Bacteriophage and Its Clinical Applications, Tr. by George H. Smith, Springfield, Ill., Charles C. Thomas, 1930.

d'Herelle, F., and Hauduroy, P.: Sur les charactères des symbiose bactérie-bactériophage, Compt. rend. Soc. de biol. **94**:1288, 1925.

Hindle, E.: On the Life Cycle of Spirochaeta Gallinarum; Preliminary Note, Parasitology **4**:463, 1911.

Hoder, F., and Suzuki, K.: Ueber die Gewinnung von Bakteriophagen aus Pankreasextrakten Centralbl. f. Bakteriol., I, O. **98**:433, 1926.

Hort, E. C.: (a) Epidemic Cerebro-Spinal Fever; the Place of the Meningococcus in Its Etiology, Brit. M. J. **1**:156, 1916.

(b) Studies in the Pleomorphism in Typhus and Other Diseases, J. Roy. Micr. Soc., 1916, no. 6, p. 528.

(c) Morphological Studies in the Life History of Bacteria, Proc. Roy. Soc., London **89**:468, 1916.

Israël, J.: Neue Beobachtungen auf dem Gebiete der Mukosen des Menschen, Virchows Arch. f. path. Anat. **74**:15, 1878.

Jones, D. H.: A Morphological and Cultural Study of Some Azotobacter, Centralbl. f. Bakteriol., II **38**:14, 1913.

Further Studies on the Growth Cycle of Azotobacter, J. Bact. **5**:325, 1920.

Klebs, E.: Weitere Beitrage zur Geschichte der Tuberkulose, Arch. f. exper. Path. u. Pharmakol. **7**:1, 1883.

Koch, Robert: Untersuchungen über die Aetiologie der Wundinfektionskrankheit, Leipzig, 1878.

Verfahren zur Untersuchungen, zum Conservieren und Photographieren der Bakterien, in Cohn, F.: Untersuchungen über Bakterien, VI, Beitr. z. Biol. d. Pflanz. **2**:399, 1877.

Die Aetiologie der Tuberkulose, Berl. klin. Wchnschr. **19**:221, 1882.

——and Mellon, Ralph: The Biological and Clinical Significance of Diphtheriods in the Blood Stream, J. Bact. **19**:25, 1930.

Krienitz, W.: Ueber morphologische Veränderungen an Spirochaeten, Centralbl. f. Bakteriol., I, O. **42**:43, 1906.

Kühn, P.: Weitere Einblick in die Entwicklung des A Formen (Pettenkoferiaformen), Centralbl. f. Bakteriol., I, O. **93**:280, 1924.

Lankester, E. R.: On a Peach-Colored Bacterium—Bacterium Rubescens, n. sp., Quart. J. Micr. Sc. **13**:408, 1873.

Lehmann, K. B.: Ueber die Sporenbildung bei Milzbrand, München. med. Wchnschr. **34**:485, 1887.

Leishmann, W. B.: Preliminary Notes on Experiments in Connection with the Transmission of Tick Fever, J. Roy. Army M. Corps **12**:123, 1909.

Leuriaux, C., and Getts, V.: Culture du Treponema pallida de Schaudinn, Centralbl. f. Bakteriol., I, O. **41**:684, 1906.

Levaditi: L'état du virus de la fièvre récurrent (Sp. duttoni) dans l'encephale de la souris, Compt. rend. Soc. de biol. **100**:1121, 1929.

Löhnis, F.: Studies on the Life Cycles of the Bacteria, Nat. Acad. Sc. **16**:1, 1921, with plates A to S and I to XXIII.

Life History of Azotobacter, J. Agric. Research **23**:401, 1923.

——and Smith, E. K.: Life Cycles of Bacteria, J. Agric. Research **18**:675, 1916.

Lourens, L. F.: Untersuchungen über die Filtrierbarkeit der Schweinepestbacillen, Centralbl. f. Bakteriol., I, O. **44**:420 and 504, 1907.

Mac Fadyean, J.: The Morphology of the Actinomyces, Brit. M. J. **1**:1339, 1889.

Magnin, A., and Sternberg, G. M.: Bacteria, New York, 1884; cited from Löhnis (1921).

Manouélian, Y.: Gommes syphilitiques et formes anomales du tréponème. Ultravirus syphilitique, Compt. rend. Soc. de biol. **104**:249, 1930.

Meirowsky, E.: Studien über die Fortpflanzung von Bakterien, Spirillen und Spirochaeten, Berlin, Julius Springer, 1914.

Mellon, Ralph: Contribution to the Bacteriology of a Fusospirillary Organism with Special Reference to Its Life History, J. Bact. **4**:505, 1919.

Life Cycles of the Bacteria and Their Possible Relation to Pathology, Am. J. M. Sc. **159**:874, 1920.

The Infectivity and Virulence of a Filtrable Stage in the Life History of B. Fusiformis, J. Bact. **12**:279, 1926.

——and Jost, Elizabeth: Filtration Experiments with the Granular Form of the Tubercle Bacillus, Am. Rev. Tuberc. **19**:483, 1929; J. Bact. **17**:56, 1929.

Miehe, H.: Sind ultramicroskopische Organismen in der Nature verbreitet? Biol. Zentralbl. **43**:1, 1923; cited from Enderlein.

Migula, W.: System der Bakterien, Jena, G. Fischer, 1897, vol. 1; 1900, vol. 2.

Morin, H., and Valtis, J.: Sur la filtration du bacille de Johne à travers les bougies Chamberland L², Compt. rend. Soc. de biol. **94**:39, 1926.

Mudd, S., in Rivers, T. M.: Filtrable Viruses, Baltimore, Williams & Wilkins Company, 1928, pp. 55-94.

Neisser, A.: Versuche über Sporenbildung bei Xerose-Bacillen, Streptokokken und Choleraspirillen, Ztschr. f. Hyg. u. Infektionskrankh. **4**:165, 1888.

Neufeld, F.: Variability of Bacteria, de Lamar Lectures, 1926-1927, Baltimore, 1928.

Ueber die Agglutination der Pneumokokken und über die Theorien der Agglutination, Ztschr. f. Hyg. u. Infektionskrankh. **40**:54, 1902.

Nicolle, C.: Sur la nature des virus invisible. Origin microbienne des Inframicrobes, Arch. Inst. Pasteur de Tunis **14**:105, 1925.

Sur les races du pneumocoque, avec remarques géneral sur les antigènes, Bull. Acad. de méd., Paris **81**:843, 1919.

——and Blanc, G.: Études sur la fièvre récurrent, Arch. l'Inst. Pasteur de Tunis **9**:81, 1914.

Nocard, E., and Roux, E.: Le microbe de péripneumonie, Ann. de l'Inst. Pasteur **12**:240, 1898.

Novy, F. G.: Ein neuer anaërober bacillus des malignen Oedems, Ztschr. f. Hyg. u. Infektionskrankh. **17**:209, 1894.

——and Knapp, R. E.: Studies on Spirillum Obermeiri and Related Organisms, J. Infect. Dis. **3**:291, 1906.

Nungester, W. J.: Microbic Dissociation of B. Anthracis, J. Infect. Dis. **44**:73, 1929.

Oeskov, J.: Étude sur la morphologie du virus péripneumonique, Ann. de l'Inst. Pasteur **91**:473, 1927.

Oesterle, P., and Stahl, C. A.: Untersuchungen über den Formenwechsel und die Entwicklungsformen bei Bacillus mycoides, Centralbl. f. Bakteriol. II **79**:1 1929.

Olitsky, P. K., and Gates, F. L.: J. Exper. Med. **33**:125, 361, 373 and 713, 1921

Otto, R., and Munter: Zum d'Herelleschen Phänomen, Deutsche med. Wchnschr **47**:1579, 1921.

Palante, B., and Koudriavtzeva, V.: De la filtrabilité du streptocoque, Compt rend. Soc. de biol. **96**:1218, 1927.

Panek, K., and Zakharoff, N.: Récherches sur la morphologie et la biologie de la forme filtrable du bacille tuberculeux, Compt. rend. Soc. de biol. **104**:607 1930.

Perrin, W. S.: Researches on the Life History of Trypanosoma Balbianii, Arch f. Protistenk. **7**:131, 1906.

Perty, M.: Zur Kenntnis kleinster Lebensformen, Bern, 1852; cited from Löhnis (1921).

Pfeiffer, R.: Weitere Untersuchungen über das Wesen der Choleraimmunität Ztschr. f. Hyg. u. Infektionskrankh. **18**:1, 1894.

Plantureux, E.: (a) Sur la nature de la lyse transmissible des bactéries, Compt rend. Acad. d. sc. **190**:224, 1930.
  (b) Sur la nature des "bactériolytes," agents de la lyse transmissibles, Compt rend. Soc. de biol. **103**:387, 1930.

Pryer, Roy: The Cause of Scarlet Fever, Am. J. Pub. Health **15**:847, 1925.

Radziewski, A.: Untersuchungen zur Theorie der bacteriellen Infektion, Ztschr f. Hyg. u. Infektionskrankh. **37**:1, 1901.

Ramsin, S.: Sur les formes filtrables des streptocoque et sur la natur de la toxin de Dick, Compt. rend. Soc. de biol. **94**:1010, 1926.

——and Givikovitch, M.: Transformations du streptocoque hémolytique, Compt. rend. Soc. de biol. **95**:952, 1925.

Rosenow, E. C., and Towne, E. B.: Bacteriological Observations in Experimental Poliomyelitis of Monkeys, J. M. Research **36**:175, 1917.

Roux, E.: Bactéridie charbonneuse asporogène, Ann. de l'Inst. Pasteur **4**:25, 1890.

Sédaillan and Gaumond: On the Filtrability of Streptococcus, Presse méd., Oct. 29, 1927, p. 1313.

Séguin, P.: Treponéma calligyrum et ultra-virus spirochaetique, Comp. rend Soc. de biol. **104**:247, 1930.
  Spirochaeta gallinarum et formes dites "ultra-virus," Compt. rend. Soc. de biol. **104**:836, 1930.

Smith, G. H., and Wilson, Elizabeth: Protobacterial Forms of B. Diphtheriae, J. Bact. **20**:25, 1930.

Stearn, Esther; Sturdivant, B. F., and Stearn, A. E.: The Ontogeny of an Organism Isolated from Malignant Tumors, J. Bact. **18**:227, 1929.

Togounoff, A.: Sur les élements filtrables du virus tuberculeux, Compt. rend. Soc. de biol. **97**:349, 1928.

Tunnicliffe, Ruth, and Jackson, L.: Bacillus gonidiaformans (n. sp.), hitherto undescribed organism, J. Infect. Dis. **36**:430, 1925.

Urbain, A.: Les formes filtrantes du streptocoque gourmeux, Compt. rend. Soc. de biol. **97**:1598, 1927.

Valtis, J.: Sur la filtrabilité du bacille tuberculeux à travers les bougies Chamberland, Ann. de l'Inst. Pasteur **38**:452, 1924.

Vaudremer, A.: On the Filtrability of the Tubercle Bacillus, Compt. rend. Soc. de biol. **89**:80, 1923.

Winslow, C. E. A., and Winslow, A. R.: The Systematic Relationships of the Coccaceae, New York, John Wiley & Sons, 1908.

Wolbach, S. B.: On the Filtrability and Biology of the Spirochaetes, Am. J. Trop. Dis. **2**:494, 1915.

Zettnow, E.: Ueber den Bau der Bakterien, Centralbl. f. Bakteriol. **10**:689, 1891.

Zlatogoroff, S. I.: Die Variabilität der Mikroorganismen als ein biologischer Faktor in der Pathologie und Epidemiologie, Deutsche med. Wchnschr. **50**: 1499, 1924.

De l'étiologie de la scarlatine, Compt. rend. Soc. de biol. **96**:1220, 1927.

De l'étiologie de la scarlatine, ibid. **187**:153, 1928.

Ueber die Etiologie des Scharlach. Gibt es ein filtrierbares Virus beim Scharlach? Centralbl. f. Bakteriol., I, O. **113**:97, 1929.

Zopf, W.: Entwicklungsgeschlichtliche Untersuchungen über Crenothrix polyspora, Berlin, 1879; cited from Löhnis (1921).

The Journal of Infectious Diseases, Vol. 60, No. 2 (Mar.-Apr., 1937), pp. 129-192*:

"Further Advances in the Study of Microbic Dissociation"
by
PHILIP HADLEY
Institute of Pathology,
Western Pennsylvania Hospital,
Pittsburgh, Pennsylvania, USA

*Please note: The following pages 161-224 correspond to pages 129-192, respectively, of the original published article in the Journal of Infectious Diseases. When citing the original article, please subtract 32 from the page numbers listed in this book. For example, page number 175 in this book indicates page number 143 in the original article. The page numbers in this book have been assigned consecutive to the numbers in the above article in this compilation, which terminates in number 159.

Jour. Infect. Dis., Vol. 60, No. 2

FURTHER ADVANCES IN THE STUDY OF
MICROBIC DISSOCIATION

PHILIP HADLEY

From the Institute of Pathology, Western Pennsylvania Hospital, Pittsburgh, Pennsylvania

INTRODUCTION

When Baerthlein in 1918, Arkwright in 1921 and de Kruif in the same year presented their studies on colonial variation and associated characteristics in a relatively small number of bacterial species; and when Mellon, during the years immediately preceding and following, was recording the elusive variability in cell form and structure of bacteria with special reference to pathology, it is to be doubted that any of these investigators of variation phenomena realized that his work was opening the door to a new epoch in bacteriology—dedicated to the new biology of the bacteria.

The fresh and invigorating stimulus that they helped to establish swept more quickly than could have been anticipated through the arteries of a science already languid before its time. Bacteriology was moving, but what its actual course was few could say. A well known American biologist has remarked that, as a rule, at least a decade is required for the general acceptance of any radical change in viewpoint relating to any fundamental biological conception; and that the time elapsing is greater if there exists any well-established opposition. And opposition there has been—partly of a passive type, bred of a scientific lethargy in the presence of new and unpleasantly stimulating ideas requiring mental re-orientation; partly militant, and often actuated perhaps by the desire to protect the crumbling foundations of things that had gained a semblance of truth chiefly because they had so long been believed.

But it is now nearly 15 years since the minds of thinking bacteriologists were first definitely turned into new trends of thought concerning the nature and organization of bacteria. And it is nearly a decade since I attempted, in an earlier monograph,[66] to review some of these new trends which, even at that date, had begun to become established; and to point out their highly important bearing on many fields of bacteriological thought and practice.

In still more recent years these newer conceptions have managed to hold their own against the lessening degree of academic inertia and opposition. Indeed it now appears from the number and quality of recent contributions that the movement is accelerated as years go by; and that, instead of running short of material for study, the subject is constantly opening new channels for alluring and profitable investigation. In view of these facts it seems desirable again to take inventory of the present situation with respect to dissociative variation, to ascertain from published records of recent years what new knowledge has been gained, as well as to note what, among the earlier premises, have been shown erroneous. It is also perhaps timely to appraise the value and limitations of some of the methods currently employed; and

Received for publication, June 9, 1936.

finally to examine the difficult problem of the possible biological significance of dissociative phenomena.

Since it has come about that there are few aspects of bacteriological study that, in one way or another, have not been influenced by the dissociative concept, the following review will undertake to outline some of the progress that has been made particularly in the following fields: (1) the study of colony form and cell morphology; (2) biochemical reactions; (3) immunological reactions; (4) studies on virulence and toxigenicity; (5) problems related to species characterization and classification. In view of space limitations, and the complex nature of studies on the antigenic and serological aspects of dissociation, it has seemed desirable to omit these from the present discussion, except as they touch upon other subjects considered.

Although the attack on the general problem of dissociative variation has been directed along a broad front, it will become apparent as we proceed that many of the significant contributions concern bacterial species of importance in the field of medicine; and it is here alone, thus far, that the chief applications of the new knowledge have begun to be made. There will be no attempt to emphasize these features, but it may be said in passing that they concern especially observations that can be, or have been, utilized in more exact methods of diagnosis of disease, in more rational prophylaxis; and to some slight extent in the realm of serum-therapy. In these fields it is of interest to note how quickly these new studies, having their origin only in the desire to understand more fully the nature and organization of bacterial species, have inevitably been directed against practical problems in medical bacteriology. And there exists perhaps some cause for gratification that this new field of endeavor, even so early, is not lacking in certain practical eventualities eagerly searched out by those whose first appraisal of a newborn conception is largely based on an answer to the question—What is it good for?

In concluding this introduction it should also be pointed out that economy of space makes it impossible to mention all of the valuable contributions made in recent times to the knowledge of bacterial variation; or to present as fully as could be wished the several important bearings a single piece of work may have on the general subject. It is possible to present only what I regard as a fair and average cross-section of the dissociation field and its possible relation to the whole variation problem, together with such comments as the written or photographic records appear to justify.

In attempting to accomplish this it appears of advantage, rather than to present the important data as related to individual species, in the manner of a seed catalogue, to relate the pertinent observations to special aspects of the dissociation problem; and thus to unify the subject matter with reference to the individual problems of study instead of organizing the data under the headings of numerous bacterial species whose highly eccentric qualities and behavior are our present concern. If, in the following presentation, important omissions occur—and I think it very likely they do—it must be pleaded that any complete survey of the affairs of the bacteria, recorded in countless journals and in many languages, is not a small matter for a brief and somewhat impromptu review. It is anticipated that sometime in the future the present writer will share in the preparation of a more exhaustive and detailed consideration of the entire variation problem as it concerns the bacteria; and in this it is hoped that the deficiencies inescapable in the present brief review may be rectified; and especially those related to biochemical and serological reactions.

THE CHIEF CULTURE PHASES

Regardless of what significance may be attached to the observation, recent studies have continued to demonstrate that a bacterial species, in the completeness of its development, is a kaleidoscopic thing; that the life of a specific population of bacterial elements, depending in part upon the nature of their environment, but probably more on an intrinsic directive force or mechanism, normally comprises successive or alternating stages or phases of development, each of which differs from those preceding or following in relation to many of the attributes, on the strength of which bacterial species have commonly been classified.

Whether there exists, in the order in which these phases follow one another, merely a haphazard alternation, or whether there exists an orderly trend to which might be attributed a cyclic significance, is an important problem and one which will be considered later in detail. In the meantime we may turn attention to the outstanding characteristics of the chief culture phases; and, for the moment, to those attributes related to colony form, since these, as Firtsch first pointed out nearly half a century ago, are fundamental.

Although several investigators have called attention to the multiplicity of colonial forms appearing in the life of a bacterial species and, although some of these forms presumably possess still unrecognized significance, there were two that stood out beyond others by reason of the clearness of their morphological differences and the frequency of their occurrence. They were termed the smooth (S) and the rough (R), between which there exist certain so-called intermediates (Sr, SR, sR). It has continued to appear that bacterial cultures spend much of their time in a slow alternation between these two phases.

The three-phase system.—But, in addition to the S and R cultures, it has frequently appeared in the literature of earlier years that there also existed a culture form characterized by a mucoid consistency and often by capsulated organisms—both of these characters appearing in species in which such attributes had not been commonly observed. Long ago such forms had been recognized for the anthrax bacillus, for Pasteurella and for the paratyphoid bacteria. Later they were reported for B. typhosus by Gay and Claypole,[55] Shimidsu[169] and Marassini.[99] In still more recent times in Shigella by Edwards;[38] the gas bacillus by McGaughney,[103] Roe,[156] and Stevens;[178] in B. paratyphosus by Nelson;[116] in Bact. megatherium by Knaysi;[80] in the anthrax bacillus by Nungester;[121] in Brucella by Plastridge;[140] in the Meningococcus by Rake;[144] in the influenza bacillus by Halster[73] and by Pittman;[138] in Bact. lepisepticum by Webster,[201] in B. pyocyaneus by Dahr and Kolb,[23] and in M. tetragenus by Reimann.[152]

While Davis[24] in 1912 had set apart the mucoid and capsulated Streptococcus epidemicus as an independent species, Walker[198] in 1923 reported that the mucoid aspect of this culture was only a temporary character of Str. hemolyticus. Other observations on this form will be mentioned later. Besides these instances there are of course the other outstanding examples of species such as the pneumococcus and the bacillus of Friedländer which are capsulated and mucoid in their more commonly recognized culture forms.

Since the colonies of the mucoid form are essentially smooth, there has been some tendency to speak of them as "mucoid smooth" colonies, and of the organisms as the "capsulated smooth" type—perhaps implying that the mucoid form of culture does not possess attributes sufficiently different from those of the smooth phase to justify its establishment as a distinct culture form. It is of course apparent, however, that bacteriologists agreed in setting up the S and R phases of the pneumococcus long before these were recognized as being in reality the M and S; and the differential phase characters here seemed sufficiently important, although the two forms of culture were believed to differ in no other respect than in the possession of the specific carbohydrate by the mucoid.

But there are other circumstances that might be regarded as better reasons for opposing the establishment of the mucoid as a separate culture phase. These are: (1) that a culture on a solid medium may appear highly mucoid while at the same time the individual organisms do not demonstrate well-formed capsules; and (2) that capsulated cells may appear in cultures that give indications of being in the smooth state. This is comparable with the finding, in smooth cultures, of occasional filaments that are especially characteristic of the rough. I do not, however, regard either of these circumstances as of sufficient importance to oppose the separation of the mucoid and smooth phases, since no culture phase is often found in an absolutely pure state. There are, moreover, differences in the consistency of the mucoid substance. It may be thin and watery, in which case the cells appear in stained preparations as mainly non-capsulated; while on the other hand it may be so firm and viscous that it adheres closely to the cell, where it may be revealed by staining methods. For these reasons it might be more appropriate to regard, as existing in a mucoid state, all cells which at the time are producing significant amounts of specific carbohydrate; and it seems likely that, in the long run, the cells found functioning most actively in this respect will be found in that colony and culture state that manifests, in addition, the physical characteristics of mucoid growth.

From these and similar observations, many of which have been personal, it has now become apparent that, among many well known bacterial species outside the so-called capsulated group there exist three, rather than two, culture phases that occur with such frequency and persistency that it is logical to search for the third or mucoid phase in all species whose dissociative pattern is not already recognized; and these phases now become the mucoid (M), the smooth (S) and the rough (R). It is possible that there should be added to this list a fourth—the gonidial type (G) which has now been recognized for several bacterial species.[82,166,167,82,71]

Of these phases it is still the smooth that is observed most commonly in the majority of bacterial species. Next in frequency is the rough, while the mucoid is still less frequently seen except where it is the dominant phase in nature, as also under favorable conditions of laboratory cultivation. As for the G colonies, although in recent years many small colony forms ("dwarf," "midget," etc.) have been reported, it is questionable how many of them are analogous to the G forms reported by Hadley, Delves and Klimek[67] in 1931. If a conclusion rests upon the presence of associated filtrable elements, apparently few of these minute colonies belong to the G category.

Limitations to the three-phase system.—While the chief culture phases mentioned above have been observed in many of the species that have been closely studied, it by no means follows that phases to which these designations may be appropriately applied are to be found in all species; or that, if they are found, they will be endowed with the same characters. Moreover, it may be anticipated that some species may possess phases quite different from those designated. This is indicated by many observations, including Reimann's[151,152] recent study of M. tetragenus, Roe's[156] report on the colony forms of the Welch bacillus, and the work of Long and Bliss,[93] Tunnicliff,[191] Grumbach,[61] Hoffstadt and Youmans,[71] Sherman and Safford,[166] Koser and Dienst[82] and Kopeloff[81] on a variety of species.

Christison,[17] moreover, has shown a considerable number of colony forms of Brinkerhoff's "lepra" bacillus, the fish tubercle bacillus, the tortoise bacillus of Friedmann and other acidfast species. Although this author did not employ the usual phase designations, her "variant I" of Brinkerhoff's bacillus may be recognized from her photographs as a fair smooth, and her "variant III" as an unusually perfect rough.

The multiplicity of colony forms associated with the tubercle bacillus and with many acidfast saprophytic organisms has also been shown by Petroff and Steenkin,[133] Blake and Trask,[11] also Paul,[128,129] have reported a series of colony forms regarded as intermediate between M and S (new designations) of the pneumococcus; and Eaton[37] working with the same species has described curious, autolytic, colony variants termed "phantom colonies." Several variants of the hemolytic streptococcus have been reported by Tunnicliff[189] and by Ward and Lyons.[199] Reimann[151] indicated seven different colony types of M. tetragenus, some being distinguished by chromogenesis of different sorts.

It may be anticipated that some degree of confusion will accompany the extension of the list of chief colony variants beyond the phases already established. One reason for this is the present tendency to describe and to accept as stabilized phases intermediate colony forms which must be interpreted as culture phases in the course of transition. One of the most frequent errors is to designate as rough, colonies still possessing a central papilla—which is almost invariably made up of residual components nearer to the smooth state; also of describing as smooth, colonies whose margins already manifest the beginning of an S to R transformation. Erroneous conclusions, due to the failure to distinguish clearly between the pure phases and intermediates, have appeared many times in the literature of recent years; and they have served only to obscure some of the essential facts of dissociative behavior. At the same time it must be admitted that some of the newly described variants probably deserve a place in the dissociative pattern of the species in which they appear. They cannot, however, merit a definite position until it is demonstrated that they possess sufficiently differential attributes, and that they occur with a sufficient frequency in culture development. One such possible addition, among the hemolytic streptococci, might be the diphtheroid phase (D phase).

UNIFORMITY IN USE OF PHASE DESIGNATIONS

Another situation conducive to confusion in designating certain culture phases lies in the use of the terms, smooth and rough. It involves the practice of assigning these terms, not on the basis of the physical characters of smoothness or roughness in the respective colony forms, as originally done by Arkwright, but upon the basis of virulence or some other attribute such as "sensitivity" or "resistance" to environmental conditions. It happened that, in the species first studied from the dissociative viewpoint, greatest virulence was commonly attached to the smooth phase, and nonvirulence to the rough. Because of this, certain authors came to regard virulence, rather than smoothness, as the chief criterion of the smooth culture type; and have accordingly designated their cultures as smooth if they were virulent, regardless of aspect of the colony surface. This and allied procedures have sometimes caused confusion, particularly with reference to phases of the tubercle bacillus. Birkhaug,[9] however, in his comprehensive monograph on variation in this species, rightly emphasizes the physical colony structure as the proper criterion of culture phase, and thus makes comparable dissociation in this species with that of others.

In conclusion it should again be emphasized that culture phase should be diagnosed only on the basis of the physical attributes of colonies. When this has been done, virulence, toxigenicity and other characteristics should be permitted to fall where they will; for it now seems probable that, in different species, there is no necessary correlation between virulence and the same culture phase. Either mucoid, smooth or rough forms may, upon occasion, carry the factor, or factors, for virulence depending on the species in question. It is also possible that in some species more than one phase may be virulent, as suggested by the studies of Loewenthal[90,91] on the hemolytic streptococcus. Whenever this situation appears to obtain, however, it is important to eliminate the possibility that one phase, upon inoculation, does not become transformed into another.

Phase designations of the pneumococcus.—A recent striking example of the need of uniformity in phase designations in different species concerns the dissociative pattern of the pneumococcus. The two phases, first termed S and R by Griffith, were designated at a time when only these phases were commonly recognized in species studied up to that date. With further study, however, and particularly with the recognition of the mucoid as one of the chief phases in several other species, it appeared that there existed in the pneumococcus, as in the pneumobacillus of Friedländer, a distinct incongruity, earlier called attention to by Topley and Wilson.[188] This involved the fact that what had been termed the smooth phase cells were capsulated, and that the cells designated rough were not so—nor were they distinctly bacillary nor filamentous. In 1934 Dawson[26] as a result of having discovered the actual rough phase, which I had earlier postulated as a necessary, but still missing, component of the dissociative pattern of this species, was able to bring into harmony the pattern of the pneumococcus with that of many other species known to possess a mucoid phase—the paratyphoid bacteria, for example. On these grounds, the phase previously termed the rough became the smooth; while the phase previously termed the smooth became the mucoid. The place of the actual rough, which had, up to this time, been missing in the pneumo-coccus picture, is now taken by the new variant introduced by Dawson in 1934. This revised scheme for designating the pneumococcus phases repre-sents the only logical interpretation of the observed facts. Further reports of investigation of this organism will be clearer if, henceforth, the new symbols are employed by all having occasion to write on this subject. While the new terms will perhaps cause some confusion for a time, it would be a greater misfortune to continue the employment of terms that are wholly inaccurate in their implications. Misunderstandings can probably be avoided if authors would state whether they are employing the new designations (N.D.) or the old (O.D.).

Dissociative pattern in hemolytic streptococci.—The first to apply the S-R dissociative pattern to hemolytic streptococci was Cowan who, in 1922, by a rigorous course of selection, produced what she believed to be the smooth and rough culture phases. Of these the smooth was apparently the more virulent for animals although the difference between smooth and rough in this respect was not strongly marked. In the light of more recent studies it seems possible that Cowan's "smooth" may have been the mucoid; and that her "rough" may have been the matt. At any rate, the lack of knowledge at that time regarding the existence of a mucoid phase (which is essentially smooth) as a component of the three-phase system, makes easily understand-able her possible failure to recognize it as such in her early experiments on streptococcus dissociation. That a mucoid phase was sometimes associated with streptococcus cultures was suggested, however, by several observations made both before and after Cowan's work.

Although Bordet, many years earlier, had drawn attention to the capsular state of virulent streptococci when taken freshly from the tissues of animals, it was first Davis[24] who, in 1912, presented evidence which, when properly interpreted, strongly indicated the existence of a mucoid or capsulated stage in the development of hemolytic streptococci; for he obtained at that time, from septic sore throats, a strain of capsulated streptococcus which filled all the requirements of a mucoid stage. This organism was later given species-rank as Str. epidemicus, but without an appreciation of the circumstance that it was only a developmental form of the common hemolytic strepto-coccus.

In 1928 Todd[186] reported two types of colony appearing on plates of heated blood agar seeded with materials taken from patients experiencing hemolytic streptococcal infections. These colony forms he termed the "matt" and the "glossy." He also described a colony form termed the "pseudo-glossy." While the glossy, which was originally nonvirulent and persistently remained so, is probably analogous to what is now called the smooth, the matt, which was virulent, has been more difficult to place in the dissociative pattern. While I, as also Andrewes at one time, earlier regarded it as equivalent to the rough, and while many of its characteristics—such as dull, pebbly or rough surface, stringy growth in broth, as well as an occasional tendency to become flattened—would seem to justify this allocation, such an allocation seems at the present time unjustified. This is chiefly because a different form of colony, possessing features that are more in harmony with the anticipated requirements of the rough has been produced by Dawson as well as by Faith Hadley and myself. This form commonly has its origin as an outburst from the smooth, as might be expected. Dawson, who has recently studied the matt form has concluded that it is an intermediate between the mucoid and the smooth, or possibly as an "abortive mucoid." Faith Hadley and myself have observed in numerous instances the fairly rapid transformation from matt to mucoid in cultures freshly derived from streptococcal infections. And, indeed, whether a culture fresh from tissues appears on primary cultivation on blood agar plates as matt or mucoid appears to depend in considerable degree on the constitution of the plating medium. We have also observed the sudden transformation of freshly isolated matt cultures to mucoid, following one or two passages through mice. There is considerable evidence that the mucoid represents the more virulent phase in mice. Dawson is of the opinion that Todd's "pseudoglossy" colonies probably represent the mucoid; and we ourselves are inclined to accept this view.

In 1922 Bitter[10] reported that, during a period of eight years, he had observed 30 cases regarded as being due to a hemolytic form of Str. mucosus. But, in all these cases except one, the mucoid form of the cultures was lost after a few transfers. In the light of subsequent observations it seems probable that Bitter was dealing with the mucoid phase of hemolytic streptococci. Indeed, this conclusion might have been drawn after the work of Walker[198] in 1923, since this author actually followed the course of transformations.

But to return to the question of the mucoid as one of the chief culture phases of the hemolytic streptococcus. Among the first to use a direct experimental approach to this problem, although not at that date recognizing the complete dissociative pattern, were Jennings[75] and Pilot and Davis.[131] The former gave convincing evidence that many strains of hemolytic streptococci could, by certain cultural procedures, be transformed into a mucoid state. Pilot and Davis observed that, by animal passage, ordinary forms of streptococci could be made to assume large, moist colonies not unlike Str. epidemicus.

In 1932 Loewenthal[90] demonstrated the presence of a mucoid form producing a severe epidemic among mice. This form was again mentioned by Dudgeon and Derru[33] the same year, and in 1933 Ettinger-Tulczynska[41] described the mucoid phase and studied its serological reactions. Also, in 1933, Tunnicliff[190] observed large, moist colonies of a strain of streptococcus from erysipelas, occurring after 15 passages on blood agar. Already in 1931 Tunnicliff[189] had observed capsules on virulent scarlet fever streptococci but not in avirulent strains. In the same year she[190] reported large, moist hemolytic colonies, similar to those from septic sore throats, arising in two cultures from scarlet fever. Subsequently this author[192b] pointed out that fibrinolysis was associated with the virulent and capsulated strains and that this property was lost upon transformation to what she regarded as the rough phase. Others who have dealt with this problem are Oppenheim,[124] Grumbach,[61] Ward and Lyons[199] and others, whose works will be mentioned on a later page.

Dawson and Olmstead,[26] however, were the first to attack the problem with a clear understanding of the dissociative background. In 1934 they presented a preliminary report in which the place of the mucoid phase in the dissociative pattern was clearly demonstrated. They examined the frequency of occurrence of this phase in cultures fresh from human sources and found methods for producing it from smooth phase cultures by serial growth on appropriate mediums. They also emphasized the parallel nature of the dissociants of the pneumococcus and streptococcus; and of both with the chief culture phases of other bacterial species.

At the same time it has been made clear from the studies of Grumbach,[61] Spicer,[175] Tunnicliff,[189.190] Ward and Lyons[199] and Howell and Burton,[74] to mention only a few reports, that there exist among streptococci colony forms and probably equivalent culture phases which cannot yet be brought into the present picture. The observations to date, including recent studies of our own, however, suggest that the dissociative pattern of these organisms is more elaborate than that of the pneumococcus, and indeed more complex than commonly supposed. At least the diversity of colony form and of cell morphology appears to be much greater.

Until further observations clarify the situation relating to colony form and culture phase among the hemolytic streptococci, the following conclusions appear justifiable: As in many other species the streptococcus manifests the three, and probably more, chief colony forms. The most commonly recognized are the mucoid, the smooth and the rough. In these, the colony features bear a marked resemblance to those of other species. As in the pneumococcus, the mucoid carries maximum virulence, while the smooth and rough forms are lacking in this quality. In many strains of the hemolytic streptococcus, however, there exists another clear-cut colony type—the matt, probably a transition form. This may reveal some degree of primary virulence for mice and following intraperitoneal inoculations at least, readily transforms into the mucoid phase in the peritoneal cavity and in the blood of these animals. Whether the matt or the mucoid is observed on primary isolation from patients seems to depend in some measure upon the kind of medium employed. While on the common blood agar mediums the matt appears to be more commonly obtained, neopeptone blood agar favors the appearance of the mucoid, as shown both by Dawson and ourselves. Some such circumstance probably explains the discrepancy between the reports of Todd and of Dawson relating to the phase in which the hemolytic streptococcus is most commonly obtained in primary isolation from infective processes. The surface and contour of the matt colonies can vary within wide limits—in fact to such a degree that some may be mistaken for smooth, while others, particularly if flatter, can be mistaken for roughs. The matt colonies may be most easily distinguished by the circumstance that their consistency is either "mushy" or markedly coherent, differing in this respect from the mucoid on the one hand, and from the smooth on the other. In addition we have observed that the matt colonies as a rule manifest greater hemolytic power than do the mucoid.

RELATION OF COLONY FORM TO OTHER ATTRIBUTES

From earlier observation[66] on colonial variation it appeared that each culture phase was closely correlated with other morphological, cultural and biochemical attributes, such as motility, presence of capsules, cell morphology, serological reactions, antigenic constitution, virulence, etc. In more recent times there have appeared reports suggesting that these correlations are frequently not observed; and implying that, after all, each attribute of a species is transmitted to subsequent cell generations quite independently of other attributes; and all or many of them independently of colony form or culture phase. For this reason it is important to ascertain whether these more recent observations are such as to support the general view that each characteristic of a species is subject to independent transmissibility.

While it has been assumed that the rough phase cultures of motile species were nonmotile, Arkwright[7] has mentioned both "motile rough" and "nonmotile rough" forms. From my own experience, I am still of the opinion that motility in the rough phase of the typhoid bacillus invariably indicates that the full rough character has not been attained. Nungester[121] has shown that all rough cultures of the anthrax bacillus are not virulent, and Nungester and Junge,[123] from their study of a paratyphoid-like bacillus concluded that in this case colony form was not related to other attributes of the culture. Knaysi,[80] studying the dissociation of Bact. megatherium was led to the opinion that each characteristic of a bacterial species is gained or lost separately in dissociation. Waaler[197a] observed that, in the dysentery species, a culture that was in smooth phase serologically "could possess smooth, smooth-rough or rough colonies"; and he concluded that smooth type cultures are not therefore characterized by smooth morphology, but by antigenic qualities characteristic of the smooth state. This author also observed that, although the morphological dissociations might proceed either from S to R or from R to S, the serological changes occurred from S to R only— and never came back. Many of the observations reported by Waaler in connection with the purely dissociative side of his work, are highly incongruous; and one is tempted to conclude that much of his work involved the use of cultures that were in intermediate phases, or perhaps imperfectly stabilized. This view gains some evidence from the inspection of his photographs.

McKenzie and associates[104] from a study of dissociation and reversion in the dysentery bacillus concluded that independent variation of any character might occur. They observed that a stabilized rough phase culture, derived from a known smooth, yielded four smooth reversion cultures no two of which were alike. They also differed from the parent smooth culture. A somewhat similar view has recently been presented by Reed[146] with respect to the variation of colony structures, capsulation and pigmentation in Serratia marcescens. He observed that variation might occur in any one of these characters independently of variation in the other two. The behavior of the genes was suggested as responsible for these results.

With respect to the hemolytic streptococci Dawson and Olmstead[28] have recently reported that, while the majority of virulent cultures observed by them were in the mucoid phase, some mucoid cultures were not virulent, thus suggesting that there might exist some other factor for virulence, associated with the mucoid state if highest virulence was to be attained. In some cases at least they observed that it was necessary for the M factor to be present in order to assure virulence. There appears to be in Dawson's case some superficial resemblance to the observation of Felix and Pitt[48,49] that, in the typhoid bacillus, a special factor for virulence (the "Vi" factor) was present in the most virulent cultures. This matter will be dealt with more fully in a later section.

To some bacteriologists these and other similar exceptions to the previously assumed correlations between culture phase and species attributes have seemed to imply that the general principle of such correlations has become disqualified. In reality, the number of instances in which such correlations do occur far exceeds the number of reported exceptions. The positive aspect of the case does not imply that in all bacterial species the mucoid phase will be the most virulent, or that hemolysis will be restricted to the smooth, or that some rough phase cultures may not produce toxin or form gas. But it does mean that, in any given species, certain important attributes are, as a rule, found associated consistently with a certain phase; and that the observable degree of this correlation is likely to vary with the degree of purity and stability of the culture phase under consideration. I suspect that the apparent tendency of many bacteriologists to employ cultures of mixed phase is in considerable measure responsible for the confusion that exists. On the other hand, such observations as those of Dawson and of McKenzie appear to deal with a circumstance worthy of special notice; and this will be considered on a later page.

It is also becoming apparent that, as suggested by the work of Wilson[204] on aertrycke (to be mentioned in a later section), allowance must be made for some degree of variation within each phase. One can imagine, for example, that a smooth phase culture recently derived from rough might possess characters different from those of a smooth freshly arising from the mucoid. In other words, the past history of the culture may be a matter of some importance, as has been shown by Andrewes and Christie[5] for the streptococcus.

In summarizing this section it may be said that, although certain attributes of a bacterial species may be able to vary independently of the culture phase, these instances are so infrequent as hardly to effect the broad generalization that each culture phase, when existing in a relatively pure state, is closely related to a certain group of characters; and that, when one phase has become fully transformed into another, some of these attributes are lost while new ones are gained. Stevens[178] in summarizing his valuable study of dissociation in Cl. welchii referred to the situation there observed with the statement: "It appears that . . . certain qualities are grouped together to form complexes recognized as stable variants." These stable variants, it may be added, are the chief culture phases, and their stability is relative rather than absolute.

With further reference to the observed non-correlation between culture phase and species attributes one other point may be advanced. Among the characters that have been more frequently reported as correlated with phase are the following: cell morphology, cell grouping, motility, possession of specific carbohydrates, tendency to saprophytic existence, antigenic constitution (including serological and immunological features), toxigenicity and virulence. Among the characters not so clearly correlated are: chromogenesis, (with the possible exception of pyocyanin formation), hemolysis and a considerable number of biochemical reactions—most notably, fermentations. With reference to these two possible groups the former might be regarded as more fully related to the intrinsic qualities while the latter embraces attributes perhaps more fully related to what might be termed contingent features. In such a grouping the possibility is not excluded that some attributes may seem to vary independently while others, though in a measure submissive to environmental influence with respect to the degree of manifestation, on the whole pursue their destined path as frequently in spite of their environment as because of it. It is one of the tasks of students of dissociative phenomena to distinguish between these categories, if they really exist; and then to ascertain whether the most important biological characters of bacterial species are not still consistently associated with a definite culture phase. This aspect of the dissociation problem will receive further consideration on a later page.

MANNER AND SEQUENCE OF PHASE TRANSFORMATION

In recent years much additional evidence has been presented supporting the view that the chief culture phases are inter-convertible, sometimes easily and at other times with considerable difficulty.[84,186] At present there exists no evidence that any culture phase is characterized by stability in the absolute sense, although there are some that have not appeared reversible by in vitro cultivation methods. On the other hand, whether there exist culture variants of a different order from the dissociative variants, but to which the term "permanent variant," could properly be applied, is an unsettled question, but acceptable evidence of their existence is lacking; and the same may be said of so-called bacterial "mutants."[66,188]

Accepting the fact, therefore, that interconvertibility of phases commonly obtains the interesting question arises: Can anyone of the three chief phases transform directly into any other phase, or does the direction of transformation follow a definite direction or trend? At present the question cannot be answered; but a few observations bearing upon the situation may be presented.

In a large number of species the intertransformation of S and R phases appears to have been recognized. Regarded superficially, the phenomenon appears to be no more than an "alternation of generation" and, if such, carries no necessary implication of cyclical behavior. Roe[156] for example observed that the transformations of various colony forms in Cl. welchii were characterized by "oscillations" and gave no evidence of a cycle; and many similar observations have been reported. In the smooth-rough transitions it is usually observed, as Dawson[26] has emphasized for the pneumococcus and as Almeden[2] has noted in the tubercle bacillus, that, while the S to R transition is a slow and gradual process, the reversal occurs suddenly and without the appearance of colonies of an intermediate nature. The same phenomenon has been observed by Shinn[170] and by myself in cultures of the pneumobacillus. I have also observed the direct transformation from rough to mucoid in cultures of both hemolytic and greening streptococci. These observations, taken in conjunction with the observed differences in the cytology of the respective cultures undergoing transformation, suggest that the path followed by the culture in the S to R transformation is not the same as the course taken in the R to S reversion. If this is true there appears to exist, even in this apparently simple "alternation" between S and R phases, the possibility of an elementary cycle. In this case the interpretation depends upon the cell mechanism accomplishing the transformations; and of this little is known except what Dawson and Shinn and myself have shown for the pneumococcus and pneumobacillus, respectively. And this consists largely of the observation that there exists at the time of transition a high degree of pleomorphism characterized by some fairly regularly occurring morphological structures whose significance is still unknown. In this connection, however, Mellon in his long series of publications on variation pointed out the possible functioning of several of these morphological entities, particularly the zygospores and gonidia, in relation to the variation mechanism.* At that time, however, they were not so clearly related to phase transformations as were the consecutively-developed morphological elements referred to in his more recently depicted cycle of the tubercle bacillus.[113]

Regarding the technic of observations in this field, it is probable that the mechanism of the smooth-rough transformations would sometimes be clearer if the culture changes were more commonly followed in colony cultures rather than in liquid mediums, as has usually been the case. Rough to smooth transformations on agar are much more difficult to obtain, especially if the rough

---

* In this connection Stoughton,[179] has presented a detailed study of the morphology and cytology (including reproduction and cell fusions) in Bact. malvacearum. The formation and liberation of "coccoid bodies," and their subsequent germination was described and illustrated photographically. In these reproductive changes the author observed clear evidence of a life cycle, although unfortunately none of this observations was correlated with the known culture phases, which should represent one of the most important aspects. In addition the especially detailed study of Meyn[114] on the morphology and biology of the Rausch-brand bacillus presented evidence of reproduction in this species by the formation of gonidia and by symplasms; but here again the transformations were not studied with a background of cultures of recognized phase. Both of these papers are important inasmuch as they demonstrate that bacteria can reproduce, in the exact meaning of this term, as well as multiply.

is perfect and well stabilized; but, when observed, they are much more adapted to detailed and accurate observation, partly because one can select for microscopic examination, or for further cultivation, definite culture areas, and not sample blindly as from a broth culture.

The recent recognition of the mucoid as one of the legitimate culture phases further complicates the problem of trend in culture development. Its recognition is so recent, however, that little knowledge has been gained regarding the place it occupies in the dissociative scheme. Several points are nevertheless clear from investigations on the pneumococcus and the bacillus of Friedländer, which have been more fully studied. First it is apparent that the only course of transformation out of the mucoid phase, as shown by Dawson,[26] for the pneumococcus, and as confirmed by Shinn and myself, is to the smooth. And the same has been demonstrated by us for the pneumobacillus. In addition I have observed the same limitation in many members of the paratyphoid group. This M to S transition, moreover, occurs in colonies only by means of wedge-shaped sectors formed at the margins of the mucoid colony. In M colonies of the pneumococcus, as Shinn and myself have observed, these wedges are difficult to recognize except under magnification, and they may be quickly obscured by a sudden reversal of the trend. On the floor of these translucent sectors lies, in more or less pure form, the newly generated smooth phase culture, and transfers from these areas bring it out in greater purity. These observations relating to sector formation in mucoid colonies bring the mechanism of M to S dissociation (in pneumococcus colonies) into harmony with the method of equivalent phase transformation in many other species. And it may be predicted that it will be equally true in still others in which the mucoid phase or its equivalent has not yet been discovered. I know of no instance in which the direct transformation from mucoid to rough has occurred; and this probably marks the first well recognized limitation in the trend of phase transformation—as opposed to the views of those who regard phase transitions purely haphazard in their sequences.

When the mucoid phase is the destination, rather than the point of departure of phase transition, common observation seems to indicate that the starting point is the smooth rather than the rough, although Dawson[217] has observed the rough pneumococcus transforming to the mucoid, and I have observed this transition once in the pneumobacillus. I have also observed the S to M transition in nearly all the members of the enteric group except in B. typhosus and the dysentery bacillus, but I have never seen the R to M transition in these species. When the S to M transition is observed in colonies the mechanism most commonly involves the often observed "wall formation" of new mucoid culture about the periphery of a smooth colony, as pictured for the pneumococcus by Grumbach,[61] Klumpen[79] and others. Successive marginal selections usually result in establishing a relatively pure mucoid form of growth. This change can occur in smooth cultures only if they have not progressed too far toward the R phase. The S to M "reversion" is less commonly seen than the M to S transformation.

In colony cultures the S to R transition in the majority of species studied appears to pursue the same course. It involves the formation of fan-shaped rough "outbursts" from certain points on the margin of the smooth colony, or the development of a fringe of rough growth entirely around the smooth colony. The difference in these two methods is manifestly due to the number of points at which dissociative action begins. Unlike the sectors in the M to S change, the rough outbursts rapidly extend their front beyond the smooth colony margin, so that the new growth may eventually transcend the original colony structure. These phenomena have been observed in a considerable number of species, including the enterococcus by Faith Hadley,[217] B. acidophilus,[64a,65] the pneumococcus,[26,216] the hemolytic streptococcus,[19,29,216] the bacillus of Friedländer,[170] several members of the paratyphoid-enteritidis group [216] and other species.

The phase transformations, beginning with the rough, whether occurring in liquid mediums where they have been observed most commonly, or on solid mediums where they are rarely seen, is usually to the smooth, although, as stated Dawson believed a transformation to the mucoid may occur in the pneumococcus. The only method of transformation in colonies is through the formation of smooth secondary colonies at or near the border of the rough growth, as I have noted in the pneumobacillus and other species. When the change occurs in broth cultures the manner of transformation is of course unknown. With one possible exception, I have not observed a direct transition from R to M, either in broth or on plates, unless the culture passed through S on its way. But these observations include only a few bacterial species; and many must be studied before one can conclude that a definite sequence exists in the phase transformations. Of all the possible interconversions, the M→S→R sequence seems to be the most securely placed.

Finally, a word may be said on the possible existence of similar culture phases and phase transitions in groups of micro-organisms that are somewhat related to the bacteria, and whose development may be studied in colony cultures upon solid mediums. This question as relating to various yeast species has been studied by Fabian and McCollough, and the situation existing among the fungi has been indicated to some degree by the interesting observations of Emmons on Anchorion and Trichophyton.

Fabian and McCullough,[44] studying five yeast species, demonstrated the existence of four different culture phases—the smooth, rough, gonidial and a transitional form occurring between the smooth and the rough. They reported that the form of yeast commonly mentioned in the literature is the smooth phase culture. The rough was characterized by dull, wrinkled colonies having a filamentous edge, and comprising greatly elongated cells. The G form showed cells much reduced in size and asporogenic; also characterized by producing acid, rather than alcoholic, fermentation. The colonies of this form were microscopic in size at the end of one week's incubation. The T or transitional form "consisted of highly refractile cells which produced the G form by the formation of a multitude of minute buds on the periphery of the cell." The authors were not able to culture this form. Immunological studies indicated a closer relationship between the R and G forms than between the smooth and the other two forms.

In its theoretical aspects this study by Fabian and McCullough is one of the most significant contributed to the knowledge of microbic dissociation in recent times. The almost exact duplication in yeasts of the essential features of colony variation, cell variation and biochemical variation now well established for the bacteria, serves once more to demonstrate that these transformations, far from being an expression of random variation, represent a parallel trend of culture development observable not only in diverse bacterial species but also in other minute unicellular organisms not distantly related. It is possible that the study of these colony and cell transformations among the yeasts, where the significance of the diverse cytology is better understood than in the bacteria, may eventually lead to a better understanding of the significance of manifestly analogous transformations among the bacteria themselves.*

That at least some of the aspects of colonial variation in bacterial species, especially the "sectoring" and marginal transformations, appear also in colonies of the higher fungi such as Anchorion gypseum, for example, was pointed out in a valuable paper by Emmons[39] in 1932. Although this author mentions somewhat similar observations made by others, and cites the explanations favored (mutations, saltations, recombinations, pleomorphism, etc.), he himself does not reach a conclusion other than to suggest that the observed changes could be the result of loss or of re-arrangement when a characteristic is determined by multiple factors. The existence of a genetic mechanism among bacteria has not yet been sufficiently demonstrated to aid in understanding the possible genetic basis for variation in these lower forms.

In concluding this aspect of the dissociation problem I wish to add that perhaps the most important issue at present is the direction or trend observable in the transformations from one phase to another and the nature of the cytological mechanism operative in these transformations. For, upon cumulative knowledge regarding these subjects, must rest the answer eventually given to the question of the ontogenetic significance of the phase variants. Evidence at present available strongly suggests this interpretation but it is not conclusive. While further study of the differential attributes of pure culture phases is of course also important, I do not regard it so important, at least for bacteriology itself, as a clear understanding of the direction in dissociative trends, and the cell mechanics involved in phase changes.

It is, furthermore, only by solving this problem that it may sometime become possible definitely to establish one of the chief categories of bacterial variability—namely, microbic dissociation, as represented by an independent series of variants related purely to the ontogeny of a bacterial species—in other words as stages in the development of the species-microphyte.† By so doing the way might be cleared for the study of other variation factors that appear to be of a different order from those observed in dissociative variation—but to whose influence the dissociative variants are themselves probably susceptible.

---

* The observations of Fabian and McCullough on the S and R colonies, and on the characteristic cell morphology accompanying them, have been confirmed in my laboratory with a yeast culture freshly isolated from the cervex uteri. It was isolated in the rough phase.

† In this connection see a later section on the significance of dissociative variation.

DISSOCIATION IN RELATION TO CELL MORPHOLOGY

From earlier studies it had been made clear that the smooth and rough phases were set apart by marked differences in respect to cell form, those of the smooth (in the Bacteriaceae) tending to be shorter and those of the rough phase longer, even to the extent of tangled masses of filaments. There relationships have been found to hold true in rough and smooth cultures studied in more recent times. It has been observed, for example, in the following species: Cl. welchii;[156,103,178,148] Salmonella pullorum;[141] Br. meletensis;[97,101] H. influenzae;[73,171,138,84] L. acidophilus;[65,81] pneumococcus;[26,216] streptococcus;[29,19,191,216] Friedländer's bacillus;[1,170,216] tubercle bacillus;[9,2,216] Dip. rubeolae;[191] Bact. fusiformis;[192] Bact. aertrycke;[204] the diphtheria bacillus[115] and for species of yeasts by Fabian and McCullough.[44,216]

Stevens[178] referred to the gradually increasing length of the cells of Cl. welchii in passing from the mucoid, through the smooth, to the rough phase. Christison,[17] however, indicated that in some of her rough phase cultures of the tubercle bacillus the cells were shorter than those of the smooth. Koser and Styron[83] found that the cells from rough and smooth phase colonies of the Sonne dysentery bacillus "presented much the same appearance in size and shape." That bacillary forms and even filaments may appear in the actual rough phase cultures of the pneumococcus has been demonstrated by Dawson[26] and by Shinn and myself. The same appears to be true for the gonococcus[145] and has been commonly recognized for the rough phase cultures of many greening streptococci. Regarding the staphylococci there is still some uncertainty.

The morphology of cells of the mucoid phase in different species still presents a problem. While in the M pneumococcus the cells are commonly capsulated diplococci, Dawson[26] has shown a considerable pleomorphism appearing at the time of transition to the smooth phase. The M cells of Friedländer's bacillus are commonly capsulated short rods of cocco-bacilli but Shinn[170] observed increased pleomorphism at the period of transition. But longer forms, and even chains of cells as in Pneumococcus mucosus, frequently manifest capsules. The matter is in need of further study.

In this connection it may be noted that Edwards[38] working with Shigella, has reported what was termed a "rough-mucoid" form. He also observed that chains and filaments were more commonly present in the smooth than in the rough. The rough cells were not capsulated but produced a mucoid intercellular substance in which the organisms were embedded. While the smooth culture gave homogeneous clouding in broth, the rough formed a pellicle and deposit. Subsequently homogeneous clouding occurred. The rough-mucoid produced a much greater amount of specific carbohydrate. Both forms were regarded as virulent for young horses.

These results depart so widely from those commonly observed in other species, with special reference to the features of cell morphology and mucoid nature of the rough forms, that it seems necessary to conclude, either that Shigella is an exception to the usual findings, or that the mucoid-rough culture of Edwards was not a pure phase culture. That the latter may have been true is suggested by (1) the nature of the growth in broth tubes, (2) the appearance of the colonies, (3) the circumstance that the rough mucoid could be perpetuated only by continuous colony selection. It might be anticipated that Shigella will eventually reveal not only a smooth phase such as Edwards describes; but in addition, a definite mucoid and a clear-cut rough. If, however, the dissociative pattern revealed by Edwards, apparently anomalous though it is, is confirmed in related species, the results will be of considerable interest.

It is often reported that certain cultures believed to be in one or another phase manifest cells marked by a considerable heterogeneity of form rather than a preponderance of cells having a morphology typical of the phase. This for example has been noted by O'Neil[125] with reference to capsulated species. And, as mentioned above, the same was observed by Dawson and by Shinn. The period of greatest pleomorphism appeared to be at the moment of transition from the M to the S, or from the S to the R phase. It therefore seems probable that in many other instances of reported pleomorphism of unusual degree the heterogeneity is attributable to the functioning of some still unrecognized transition mechanism between the phases. A similar situation was observed by Hort many years ago, and referred to by him as the "reproductive explosion" always heralding a change in the variation trend. This view is also in harmony with the hypothesis of Mellon regarding the mechanism through which certain transitions are accomplished.

STABILITY OR REVERSION AMONG THE CHIEF CULTURE PHASES

While the relative instability of smooth cultures has been commonly recognized, opinions have differed regarding the stability of the rough forms, especially on the nature and results of their reversion to smooth. While it has been common experience that reversion ultimately occurs, it is often only with great difficulty. The degree of difficulty seems to depend on the extent to which the rough character has become "fixed"—which, in turn, probably means the degree of advancement and of purity of the rough elements. Dulaney[34] for example reported that the "early" rough phase of B. coli could be transformed easily to the smooth, while the reversal of older and better stabilized rough forms was difficult. This observation reflects common experience.

Working with four "phases" of B. pertussis, two of which were apparently in the smooth range, and two in the rough, Leslie and Gardiner[86] in 1931 reported that "phase IV" (which appeared to be the most advanced rough) in some strains became "firmly and irrevocably established." All other strains were found inter-convertible by some means. Their "phase III" was a partial rough (intermediate) and could transform further to IV or revert to I. Their "phase IV" may therefore be regarded as an example of a fairly perfect rough, comparable perhaps with the rough of the influenza bacillus described by Kun and Fenyvessy,[84] next to be considered.

In the most important recent study of dissociation of the influenza bacillus, including the authors' views on the possible epidemiological significance of the observed variants, Kun and Fenyvessy in 1932 reported details on both smooth and rough phases together with intermediates. The S to R, and some of the R to S transformations were produced easily, but their extreme rough cultures, characterized by thread forms and twisted mycelium, in some strains remained unchanged for years; and their reversion to smooth could not be accomplished by use of any of the mediums most favorable for the propagation of smooth phase cultures. Indeed it was only by animal inoculation that the most stable roughs could be converted; and even by this method only partial or transient success was sometimes attained. By both intraperitoneal and subcutaneous injections the filaments could be brought back to the smooth coccobacilli, but the results obtained by the former method were only transient, since reversal to the filamentous state occurred after a few passages on chocolate agar. When, however, the injection of the filamentous roughs was made into mice by the subcutaneous route and some degree of abscess formation followed, the smooth form isolated from these areas was considerably more stable; and in some of the cultures the original virulence was regained. It is quite natural that in commenting upon their results the authors take opportunity to point out the possible bearing of their work on the epidemiology of influenza.

Although it might be anticipated that smooth cultures derived from rough would be identical with the smooth from which the rough was derived, this has often not been the case; and it is possible that observations of this nature possess a bearing on one of the most important aspects of dissociative behavior, as will be considered later. White[203] noted that in B. paratyphosus B, while morphological reversion from rough to smooth readily occurred it was not accompanied by a replacement of the smooth, stable antigen. Waaler[197a] observed in cultures of the dysentery bacillus a morphological but not a serological reversion occurring in the rough to smooth transition. Burnet,[14] however, reported characteristic smooth cultures of B. enteritidis upon reversion from the rough, and Koser and Styron[83] obtained from rough to smooth transformations in the Sonne dysentery bacillus smooth forms that "closely resembled" the parent smooth. Almeden[2] noted in a culture of the avian tubercle bacillus that a smooth phase recovered from the rough might not possess the virulence of the original smooth line. Dawson[25] and others have shown that, under certain conditions of cultivation, a smooth phase (N.D.) pneumococcus may be caused to revert in vitro to a smooth differing in serological type from that of the original smooth. McKenzie and associates[104] reported that when a derived rough of the dysentery bacillus reverted to smooth there appeared in the progeny from separate colonies cultures that differed from each other, and also from the original smooth, in several recognized attributes. Other instances could be mentioned, but those given above are sufficiently illustrative.

While the foregoing observations relating to new forms of smooth culture arising on the reversion from the rough cannot at present be satisfactorily explained, and although, in a general way, smooth cultures recovered from roughs possess, or quickly attain, the attributes of the original smooth type, there is a growing possibility that, in the cytological basis underlying the rough to smooth transition, there exists a cellular mechanism for kinds of variability not recognized in the present dissociative scheme, but in reality a part of it and dependent on variability factors of a different order from those that determine the superficial morphologic attributes of what I regard as the ontogenic variants or culture phases.

If such a cytological mechanism exists, particularly in rough phase cultures on the threshold of entering the smooth phase, it is further possible that it involves the functioning of gonidia (or perhaps endospores in species that produce them), as has been advanced by Mellon* in several publications, and demonstrated by Cunningham.[22] While no conclusions are justifiable at present it may be pointed out that, if such differences as have been observed in the attributes of reverted smooths are actually dependent on some transition "Anlagen," these must be present, and function most strongly, at some point in the transition from rough to smooth phase. And this, in turn, suggests the importance of searching for further evidence of a cellular mechanism responsible for S to R transformation. It has seemed probable, as I[68] have indicated earlier, that if any special function attaches to the rough phase it is likely to be a reproductive one and concerns, at least in part, the formation and liberation of gonidia. Complementing this circumstance Mellon has for a considerable time pointed out the probable part played by gonidial bodies, as well as endospores, as transition anlagen operative in at least some of the phenomena of bacterial variability.* If both of these hypotheses were true, there would be provided an adequate explanation of those instances in which it appears that variability in certain attributes comes especially into evidence in those smooth phase cultures recently derived from a converted rough. And it may be noted in passing that it is perhaps not without significance that the transformations of the serological types of pneumococcus, observed in vivo and in vitro by Dawson and others, have been accomplished only under experimental conditions in which the original mucoid culture of known serological type, to be transformed, had first been brought at least closer, to the rough phase (N.D.).

While it is freely admitted that these considerations are at present highly speculative, they have been derived from the observation of dissociative behavior in many bacterial species, and have a sufficient basis in fact to suggest the desirability of further study of the following aspects of the problem: (1) The further demonstration in as many species as possible that a significant degree of variability exists among single-colony smooth phase cultures derived from converted roughs of known ancestry; (2) further attempts to ascertain whether the transition from a perfect rough to smooth phase involves the production, liberation and germination of gonidia; (3) to ascertain whether different gonidial elements carry potentially the charactersitics that are later to mark the smooth phase cultures derived from them. If these questions can receive a definite answer in the next dozen years I believe it will be a sign that bacteriology is advancing.

In concluding this section a word may be added regarding the qualifications of any colony or culture form that is entitled to designation as a rough. Reports of recent years serve to emphasize the conclusions reached at the

---

* Mellon's views on this subject were crystallized in a still unpublished paper presented at a symposium on variation and filtrability at the Chicago meetings of the Society of American Bacteriologists in 1934.

time of my earlier review—that among rough phase cultures in general there exist many varying degrees of roughness; and that the results of any study of the morphology, the biochemistry, the serology, the immunology or the virulence of such cultures will depend, first of all, on the degree of purity or of degree of advancement of the rough phase. To state that rough and smooth phase cultures of a given species possess a similar cell morphology, or similar biochemical or serological reactions is unquestionably to indicate that the actual status of the culture forms was not correctly diagnosed, or that the phase was so slightly stabilized that transformation occurred during the test. Among the 500 or more publications during the past five years, dealing with the attributes of rough and smooth cultures, their interconvertibility and similar matters, it is distinctly the minority that present evidence of being concerned with well purified and stabilized culture phases. And, until bacteriologists can find ways and means of assuring themselves that their culture phases are actually what they are designated, the journals will continue to be cluttered with contributions better left unpublished. It is to be hoped, however, that as the primary aims and conditions of studies in the field of dissociative variation become more clearly defined, the period of nondescript contributions will pass, and that the more definite and important problems will be undertaken. In the meantime, I would recommend to all those verging on their first publication in the field of microbic dissociation to first read two or three papers whose subject may be in quite a different field from that of their own research, but which are examples of a keen and careful study of the factors involved. These are listed in my bibliography.[204, 26, 84] Indeed, they merit a second reading.

FILTRABLE FORMS OF BACTERIA

From the viewpoint of the majority of American bacteriologists the question of the existence of special filtrable forms (that is, excluding "dwarfed organisms," "bacterial fragments," etc.) appears to be still in the balance, although favorable attitudes have occasionally been expressed,[215] This is in marked contrast to the situation in Europe where the filtration studies, dealing primarily with members of the acidfast group, have for the most part been accepted. In many of the instances reported in the following pages the filtration technic may be regarded as adequate and well controlled, while in other cases there has been revealed a curious disregard of several of the primary conditions, only upon whose fulfillment could successful results have been anticipated.

The tubercle bacillus and other members of the acidfast group.—Following the numerous reports of successful tests with sputum, milk, blood, exudates, tissue suspensions, etc., as well as of cultures of the organisms, reported chiefly by the French School, Sergent and Pribriano[165] presented substantial evidence by filtration technic of the presence of a filtrable form of the tubercle bacillus in the atrophic cirrhosis of Laennec. In his work the results of guinea pig inoculation with filtered and unfiltered ascitic fluid were identical, although no organisms could be demonstrated in the materials.

Panek and Zakharoff[127] reported in detail a convincing study on the filtration of the tubercle bacillus and on the morphology and biology of the filtrable forms. The filtrates were derived from the suspensions of organs of guinea pigs previously inoculated with the bacillus of Koch, then with a special product known as "tuberculotensine"—a tuberculin-like product. From the filtrates were derived by special cultivation methods cocci grouped in diplo- and tetrad-forms, which later gave way to bacillary and filamentous structures. These, in turn, were followed by acidfast and non-acidfast bacillary forms characteristic of the original culture. Broth cultures of the granular form of the organism, at the age of 24 and 48 hours, gave the best results on filtration (75% positive), while filtration of cultures at the age of seven to 10 days rarely afforded positive results.

In 1931 Mellon[107] reported the existence of filtrable forms of a grass bacillus, the culture filtered being characterized by the presence of non-acidfast granules. In the filtrates these appeared to undergo a slow germination into a non-acidfast diphtheroid, from which the original culture was again obtained. In 1932 Mellon, Richardson and Fisher[113] demonstrated filtrable forms in the avian tubercle bacillus, using cultures characterized by granule formation. They also presented what they regarded as the life cycle of this organism, comprising four stabilized stages and three transitional stages, the second (gonidial phase), being reported filtrable. The life cycle mentioned is considered in greater detail on a later page of this review. The filtration studies of Mellon and his associates have concerned themselves in nearly all cases with cultures in which the gonidial granules have been present in considerable numbers, and their interpretations are based on the germination of these bodies and their slow evolution into the common bacillary form. In this way they allocate the filtrable form to a definite place in the cycle of the tubercle bacillus.

Fejgin[46] has reported successful results in the filtration of the tubercle bacillus, but not so in the case of Pinner and Voldrich[136] working with human strains. They observed that the presence of acidfast organisms from the air could make difficult the appraisal of culture plating, particularly perhaps in desert regions where their work was performed. Cantacuzene,[15] however, has reported obtaining filtrable forms of the "leprosy bacillus" from cultures.

Streptococcus.—On the filtration of organisms from measles reports have been made by Duval and Luzenberg[35] and Tunnicliff.[191] The former doubted the existence of special, filtrable forms but believed that the smaller coccal elements in the blood of patients were filtrable, while the larger forms present in cultures were not. Working culturally with the measles diplococcus Tunnicliff observed minute colony forms suggesting the G phase, and stated that some of the elements present in cultures passed Berkefeld V but not N candles. Evans,[43] as well as Rosenow[157, 158] in many of his papers on encephalitis and poliomyelitis, has presented further data on the filtrability of cultures from encephalitis. Evans was inclined to regard the gonidial bodies as the actually filtrable elements. Rosenow's more recent studies will be presented subsequently. Richardson and Mellon[155] working with one of Rosenow's strains from encephalitis reported successful filtration results. They submitted filtrates derived from peritoneal washings of previously inoculated rabbits, also from brain emulsions, to the Hauduroy technic, with the result that the original streptococcus was eventually recovered. Usually on the third to sixth plate in series there arose minute colonies of pleomorphic diphtheroids which, in turn, transformed into a greening streptococcus resembling the original strain. The original filtrates, inoculated into monkeys, produced no apparent lesions. In discussing their results the authors were not inclined to regard the filtrable forms in any sense related to the "virus" of encephalitis. These results serve to recall the earlier (1928) study of Fei-fang Tang and Castenada[47] in which diphtheroids were recovered three times from filtrates of brain emulsions from rabbit encephalitis. Also the more recent work of Prissick[143] in which she obtained diphtheroids as one stage in the upgrowth from filtrates of cultures of a green streptococcus, as will next be described.

Two of the most important reports on filtration of the streptococcus are those of Prissick and of McKinney,[105] both in 1933 and both dealing with the green streptococcus. Prissick grew her cultures to be filtered in Kendall's medium. Berkefeld N and Chamberland L[3] and L[5] candles were employed. In 20 samples of filtrate no growth appeared except in the form of cultures that appeared to be biologically related to the culture of origin. Usually the first growth appearing on plates, or in tubes of K medium seeded with the filtrates, did not appear for several days; and final recovery of the culture was not attained until four to five weeks had elapsed. On plates submitted to the Hauduroy technic growth often did not become fully apparent until the lapse of about two weeks. The well planned control tests constituted an important part of her technic. After a period in which the growth possessed no definite structure aside from granules, diphtheroid forms appeared, followed by recognizable streptococcal elements. Four stages in the evolution of the streptococcus were recognized.

It is perhaps noteworthy that Prissick, like Richardson and Mellon mentioned earlier, observed a diphtheroid phase of the organism emerging as one of the first recognizable culture entities. Although it is true that "diphtheroids" of one sort or another represent common laboratory contaminants, it is also becoming clear that a diphtheroid stage is present in the life history of many streptococcus, both of the hemolytic and of the greening types.

Another valuable contribution to studies on the filtrability of the green streptococci is that of McKinney[105] in 1933. In addition to a detailed study of the morphological types present in dissociating cultures, and of their biochemical and serological differences, this author studied the filtrability of the different phases. She took the desirable precaution to examine her cultures for the presence of minute coccus forms and granular bodies which suggested filtrable elements. McKinney stated that cultures which retained their "normal" streptococcus morphology up to the moment of filtration gave only sterile filtrates. Only those filtrates were positive that were derived from broth cultures in which, after a growth period of 96 to 144 hours, the culture had been reduced to an amorphous mass in which minute cocci and granules could be recognized microscopically.

Certain details of the author's technic merit restatement. The filtrates (including fresh broth added to the cultures just before filtration), derived from Seitz filters of the Manteufel model, were incubated in the original tubes and not distributed nor inoculated into other mediums. Only after the first growth had appeared as a faint precipitate on the walls of the tubes were they first opened and the growth substance transferred to plates. Visible colonies first appeared after five or six passages by the Hauduroy method. The first evidence of growth on the plates was a slight greening of the blood. When growth was fully established the "yellow bleaching" effect was observed as in the parent strain. Usually eight to 12 days was required for upgrowth in the filtrate tubes; and the first visible organisms did not resemble the parent strain either morphologically or biochemically. Eventually, however, the organisms transformed to streptococci "with attributes similar to, but not identical with, those of the original streptococcus strains." Control cultures, filtered at the same time, never passed the filters.

As related to studies on the filtrable forms of streptococcus should be mentioned the recent work of Rosenow[158] on the relation of streptococci to the viruses of encephalitis and poliomyelitis, following his earlier report on the streptococcus origin of the virus of fox encephalitis. By growing streptococci in a new medium (autoclaved chick mash) there were prepared inoculums which, upon inoculation of animals, produced symptoms and lesions characteristic of the infections mentioned. By use of filtrates of brain and cord emulsions of these animals he was able to produce in monkeys symptoms and lesions characteristic of the action of the "natural" virus. After 26 or more transfers the experimentally produced viruses "became thoroughly established and behaved quite like the 'natural' viruses of these diseases." Each of four strains was passed three times through the streptococcus-to-virus phase and through the virus-to-streptococcus phase. These viruses were obtained from streptococcus strains of various origins.

The question of the possible relationship of filtrable forms of bacteria to the so-called viruses is unquestionably one of the most important problems in bacteriology today. Considering the great diversity of the virus group, and the considerable resistance on the part of bacteriologists, even to the acceptance of filtrable forms of bacteria, it is likely that many years will pass before a final answer to the question will be possible. On the other hand, there is no special evidence to indicate that all viruses belong to the same category or that all are of the same essential nature. There is no more reason for believing that some viruses may not originate from bacteria than for assuming that they have some other origin—or that they are, from beginning to end, self-perpetuating entities. Indeed there is some evidence that suggests that some of the viruses are the offshoot of some more fundamental supporting mechamism. But, as Bordet once remarked, it is sometimes more difficult to prove that a thing is not true than to prove that it is; and the evidence that Rosenow has most recently submitted regarding the relation of streptococci to at least one variety of virus can no longer be lightly disregarded. Considering the nature of his evidence, it requires not much more imagination to believe that he has produced an encephalitic virus than to believe that he has created a new disease. And it appears to be one or the other.

Spore-forming aerobes.—Among other bacterial species in which the problem of the filtrable forms of bacteria has been studied are Bact. megatherium, reported by Rettiger and Gillespie,[153] and by Knaysi,[80] B. mesentericus and B. vulgatus by Flynn and Rettger[53] and B. mycoides by Lewis.[87] All results were negative.

Rettger and Gillespie obtained certain small colony forms that may have been related to the G phase and suspensions of these cultures were submitted to filtration; and the filtrates to "washing-forward experiments." A considerable number of the bottles and the majority of the Petri plates revealed the growth of cultures that were regarded as contaminants, and no indication of the existence of filtrable forms was detected. Experiments on the filtrability of Br. abortus and Br. melitensis were simliarly negative. The authors also reported their inability to demonstrate the presence of filtrable forms in raw sewage. Knaysi also reported an unsuccessful attempt to demonstrate filtrable forms of Bact. megatherium.

Flynn and Rettger, working with B. mesentericus and B. vulgatus, employed for filtration broth cultures of what they believed to be the G form, and also cultures lysed by the bacteriophage. Filtrates held for a considerable time showed no growth. In discussing their results the authors expressed the view that the G colonies were those derived from cells that were "relatively weak and inactive." Aside from reported "filtrable forms" that are manifestly contaminants, the authors were inclined to regard all other instances as explainable on the grounds of passage through quite coarse filters of a few organisms, especially if very small. These presumably constitute the so-called "filtrable forms" of bacteria.

B. mycoides was briefly studied by Lewis in connection with an inquiry regarding dissociation and life cycles in this species. He centrifuged broth cultures until relatively clear and then passed them through Berkefeld V, N and W candles. The filtrates were distributed and then held at room temperature. Subsequently, cultivation on agar and in broth resulted in failure. These results are in harmony with the failure of Lewis to find evidence supporting the theory of reproduction by means of gonidia, and are entirely at variance with the results of an earlier comprehensive study of the same problems by Oesterle and Stahl.[124]

Other bacterial species.—In 1932, in relation to a detailed study of the dissociation of Staph. aureus, Hoffstadt and Youmans[71] studied the filtrability of G phase cultures and concluded that certain elements in these cultures were filtrable. A repetition of these experiments, however, led the same authors in 1934 to report that neither their SY nor G cultures produced filtrable forms.[72] Swingle[182] also reported an unsuccessful attempt to demonstrate the filtrability of G forms of cultures of Staph. aureus. The infrequency of occurrence of the G colonies led the author to doubt that they represented an essential stage in the life cycle of this species.

In 1932 Almon and Baldwin[3] presented their results on the variation and filtrability of cultures of Rhizobium, derived from clover and pea nodules. Thirty-two filtrates through Berkefeld N candles from cultures lysed and cultures not lysed were subjected to the Hauduroy technic, five days usually elapsing between plate transfers. When growth appeared on the plates it occurred usually by the seventh transfer, and contaminations were not reported as causing undue difficulty, thus suggesting a technic that was adequate. All of the filtrates but one yielded culture growth of some variety. In three cases the original culture form was isolated from the filtrates, while the remaining cases involved "aberrant forms" of Rhizobium. By agglutination and other tests it was shown that at least some of the aberrant types were related to each other, as well as to the original strain. The work of these authors suggests, but does not prove, that the aberrant forms arose from the filtrates as a result of the germination of filtrable elements. This concerns, and is in harmony with, a point stressed by Mellon in many of his publications dealing with the possible functioning of gonidial bodies as a mechanism of bacterial variability, and of possible importance in certain broader aspects of variation, not strictly related to the culture phases. Even if one adopted the view that the five variants reported by the authors existed in small numbers in the parent culture, as cells of common morphology (i.e., as opposed to gonidial bodies), it would still appear that that, as a group, they possessed different filtrable attributes; and so permit one to regard the filter candle as a selective mechanism in the sense in which Mellon has dealt with this problem.

With reference to the Shiga dysentery bacillus, all attempts to confirm the experiments of Hadley, Delves and Klimek[67] have resulted in failure; and this appears to be true also for filtration tests with the G forms of the majority of other bacterial species. Sanarelli and Allesandrini in 1931 reported negative results by the use of their collodion sac method on the dysentery bacillus although successful with the bacillus of typhoid. In 1933 Wycoff[208] reported failure in attempts to filter the Shiga bacillus. In those cases where there occurred a rapid upgrowth in the filtrate tubes (24 to 72 hours) organisms of the original form were found in the filtrate within one hour of the time of filtration, thus explaining the sudden reversions. The forms usually found were the "dwarfed" forms which he regarded as characteristic of the growth in Kendall's medium and lithium chloride broth. In Wycoff's "slow reversions" the organisms present were regarded as contaminations.

Dienst[31] in 1933 and Koser and Dienst[82] in 1934 reported on a small colony variant of the Sonne dysentery bacillus. Although a rare form in the experience of these writers, they regarded it as analogous to the G form of the Shiga bacillus. Filtrates of their cultures, however, through Seitz filters, handled in a variety of ways that appear to be free from criticism, gave only negative results. These authors were not inclined to regard the G forms as comprising a special culture phase but as a slowly growing variant that had lost its "normal growth vigor."

In 1931 Kendall[78] recorded the successful filtration of bacterial variants derived from the growth of the common form of the typhoid bacillus and other species in his special medium (K medium). When filtrates were inoculated into the same medium visible growth appeared in 24 hours and the original form of culture was quickly derived. As I[68] have suggested earlier, this rather sudden recovery of the original culture is ordinarily not characteristic of filtrable forms. Although Mellon reported success in the use of a considerably modified K medium in producing filtrable forms of the tubercle bacillus, and Prissick made use of this medium in her filtration studies with the greening streptococcus, Kendall's results in general have not been confirmed.[164,21,16] Varney and Bronfenbrenner[196] believed it probable that the nature of the K medium was such as to render more permeable the pores of filter candles thus permitting the common forms of the culture to pass.

Sanarelli and Allesandrini[162] by the use of collodion sacs were able to demonstrate the passage of filtrable forms of both typhoid and paratyphoid (B. icteroides) bacilli. In some instances this was demonstrated both in vitro and in vivo. In this work special forms of culture were encountered highly suggestive of the G forms of the Shiga bacillus, both in colony form and morphology of the cells.

In a recent study devoted to the dissociation and filtration of B. acidophilus Kopeloff[81] reported a failure to demonstrate filtrable forms of this species. In plates inoculated with filtrates minute colonies were frequently observed. But, since these comprised diphtheroid bacilli, since their fermentation reactions were not characteristic of the original culture and since these new forms seemed to remain stable, he concluded that they represented contaminants.

Filtrable forms of bacteria in nature.—If filtrable forms of bacteria constitute a natural stage in the development of bacteria, as many suspect, it would seem probable that their occurrence in nature would be detectable. In 1931 Sherman and Safford,[166] acting on this possibility, demonstrated the presence of filtrable organisms in soils, decomposing manure, hay infusions, fresh human faeces and market milk. Sherman, Safford and Brueckner[167] later reported on the so-called "primitive bacteria" in milk and cheese. In these instances, as also in the work of Brueckner and Sherman[12] on aseptic milk, the employment of a dilution method, even without filtration, served to demonstrate the presence of minute forms in dilutions so high that all ordinary forms of bacteria had been eliminated in earlier dilutions. In many cases the primitive bacterial forms, by application of the Hauduroy technic to filtrates or to out-dilutions, were transformed sufficiently to assume the cultural features of recognized bacterial forms.

This important work of Sherman and his associates was confirmed by Roth[159] working in my laboratory in 1931–32 on the filtrable forms of bacteria in city sewage. Here also, both filtration and out-dilution methods revealed filtrable forms numerically in excess of the ordinary bacterial population. McDaniels and Neal[102] in 1932 also demonstrated the presence of filtrable forms in serial platings from Berkefeld N filtrates of polluted water. The initial growth on the plates was characterized by a delicate film. The number of gram-negative rod forms and of gram-positive granules increased with successive transfers. Rettger and Gillespie,[153] however, were unable to detect the presence of filtrable organisms in city sewage.

The results of these studies on the existence of filtrable forms in nature are often difficult to evaluate since, as Sherman himself states the issue, the work does not always answer the question whether the organisms appearing in the filtrates or in the out-dilutions are definite bacterial species, or a state or stage in the development of ordinary bacteria. While final conclusions may not be justified at present, the slow upgrowth of cultures on plates seeded with filtrates or out-dilutions, as observed by the authors mentioned above, as also by the present writer in a considerable number of still unpublished experiments dealing with bacteria, bacteriophage and sewage in river water suggest the gradual evolution of bacterial elements from filtrable forms present in the original material.

Conclusions on filtrable forms.—When one considers with what tardiness there has occurred a fairly general acceptance among bacteriologists of the primary facts of dissociative behavior and of the various culture phases— even when, as Winslow remarked, smooth and rough colonies had been staring them in the face for more than half a century, one cannot be surprised at the failure of many workers to recognize the existence of filtrable forms of bacteria, the demonstration of which is so much more difficult.

It is probably only under exceptionally favorable conditions that genuine filtrable forms can be demonstrated easily and with any degree of regularity. The original state of the culture material is the primary requisite. Because of this it would appear more important than is indicated by many recent works on filtration to possess, at the beginning of the experiments, some assurance that bacterial elements, whose morphology and size suggest their possible filtrability, are present before filtration in the material subjected to the test. Aside from this the next most important requisite is to utilize such mediums or conditions of cultivation as will insure the upgrowth of the filtrable elements, presumably the gonidia, regarding whose frequent dormancy there can be no question. In this respect the most important problem in securing the upgrowth in filtrates, as Mellon has recently emphasized, is that of breaking dormancy.

Until the vagaries of the filtration process, as well as those of the bacteria themselves, are better understood, it is difficult to shape a set of criteria or postulates on the basis of which the presence of special forms of bacteria can be recognized. As Zinsser[213] has already pointed out, it is doubtless of advantage to demand that certain conditions shall be fulfilled before one concludes that he is dealing with special filtrable elements. But, whether the conditions applicable to one bacterial species will be applicable to another is still a question. In the meantime, it appears sufficient to insist on only four conditions: (1) that the form of culture observed in the filtrates shall develop slowly; (2) that it shall differ morphologically and in other ways from the common culture type and from that submitted to filtration unless the latter is of the granular type; (3) that eventually, and by some experimental means involving preferably in vitro methods, its biological relationship to the parent culture shall be demonstrated; (4) that the common morphological types present in the original culture shall not appear in the filtrates, nor in or on any of the mediums primarily inoculated from them. These fairly simple requirements are in harmony with the conception of the identity of the filtrable forms with the gonidia, of the frequent dormancy of these seed-like elements, and of the filter candle as an isolation mechanism.

The general result of earlier observations on this aspect of dissociative behavior has been to demonstrate that, with special reference to the smooth and rough phases, it was more commonly the former that carried greater virulence and toxigenicity. To this generalization there were provisionally excepted the anthrax bacillus whose rough phase has often been shown virulent, and certain strains of hemolytic or greening streptococci.[4,186] These exceptions have not, however, been well sustained, since further experiments have demonstrated that a somewhat different situation exists in the hemolytic streptococci;[28,199] and it is now known that rough cultures of the anthrax bacillus, when introduced into the body of the animal, may transform into a mucoid state. The situation with reference to the greening streptococci will be considered on a later page.

Among the bacterial species in which only the smooth and rough phases are commonly recognized there has accumulated additional evidence that the smooth is the more virulent. This has been demonstrated in recent years for the plague bacillus,[137] for Cl. welchii,[13] for B. pseudotuberculosis rodentium,[142] for Brucella,[101,62] for B. pertussis,[168,86] for B. coli,[34] for the diphtheria bacillus,[211a,117] for B. cholerae suis,[89] for the influenza bacillus,[84,138] and for other species to be mentioned in greater detail in the following pages.

In her study of dissociation in the influenza bacillus Pittman[138] stated that her S phase cultures showed cells that were sometimes capsulated and which produced a specific carbohydrate, thus suggesting the presence of a partial, or beginning, mucoid phase, which has not yet been fully described for this species.

Dochez, Mills and Kneeland[32] observed the smooth phase of the influenza bacillus in the nasopharynx of patients during the acute stage of infection, although they did not regard this organism as instrumental in the etiology of influenza. They thought it more probable that the virus infection made the membranes more favorable for the development of this form. The important study of dissociation in the influenza bacillus, and it's possible epidemiological significance, by Kun and Fennyvessy[84] (who regarded it as the chief etiological agent) is dealt with further in the section on stability or reversion of the chief culture phases. They found the S more virulent.

An exception to the view that, in general, greater virulence commonly attaches to the smooth phase, has recently been made by Roe[156] for Cl. welchii. It was reported that all six of his variants except the "M" (perhaps equivalent to the G form)* were virulent; but that cultures which showed some degree of filament formation (suggesting perhaps the sR phase) were slightly more so. Judging from the photographs accompanying the study of McGaughney[103] on the same species, he found most virulent or toxigenic a culture type that was presumably in the rough range (pl. XXXII, fig. 4). In his fig. 10, pl. XXXIII, however, showing the rough organisms from the

---

* It should be noted here that the use of the designation, "M", as for midget, in Roe's case, for the purpose of indicating any colony type or culture phase except the mucoid, is likely to produce confusion in phase designations.

colony of "variant II," the short, coccoid form might suggest that the rough phase culture was on the point of passing to the smooth (or mucoid). This culture form, moreover, was reported to show some capsules and to produce more hemolysin than "variant I." The results, as reported, are not in harmony with expectations, nor in conformity with the observations of others.

Studies on the relation to culture phase to virulence, production of hemotoxin and antigenic structure in the Welch bacillus were conducted by Orr, Baker, and Reed[126a] and by McGaughey.[103] The former reported that both the smooth and the rough cultures were pathogenic upon intramuscular injection into pigeons, but that the minimum killing dose of the smooth was about ten times greater than that of the rough. They observed, however, that when the point of inoculation of pigeons that had died from infection with the rough cultures was examined, as high as 50% smooth forms could be recovered. They therefore conclude that "the death of the animals may result from the S types arising from an R to S dissociation in vivo." In their study of the relation of haemotoxin production to the S and R types they showed that in tube experiments the smooth cultures effected a drop of 70% in haemoglobin while the rough cultures determined a drop of only 38%. Experiments performed in vivo indicated even a greater difference between the S and R cultures. Despite these differences between the rough and the smooth forms, the authors observed that the fermentation reactions were the same. The results in general are those theoretically expected, but are to some extent at variance with those reported by McGaughey.

In the tubercle species some difficulty has been encountered in evaluating reports regarding the virulence of the various culture phases by reason of an occasional lack of agreement as to what colony form constitutes the smooth and rough phase respectively. Petroff and Steenkin[132] in 1930 demonstrated the greater virulence of the smooth phase cultures of the avian type, and the same relation was reported for bovine strains. In the later paper by Steenkin, Oatway and Petroff[177] the colony forms designated "smooth" (pl. 31, figs. 3 and 4) are quite different from the smooth colonies reported earlier, and cannot be regareed as characteristic. Reed and Rice[147] showed a correlation between colony structure, acid agglutination and virulence in a bovine strain. The colonies that were designated "smooth" (pl. 1, figs. 8 and 9) and which were reported as being more virulent than those designated "rough," cannot, however, be regarded as belonging to the smooth phase. These authors appear to have followed the unsatisfactory terminology used earlier by Petroff, who eventually changed the basis of his designations in such a way as to make the symbols S and R indicate "sensitive" and "resistant" instead of smooth and rough.

The confusing designations of colony forms in the tubercle bacillus has been recently noted by Smithburn[173] in one of a valuable series of papers dealing with variability in this organism. Making use of the symbols that have now become well established for other bacterial species, this author described colony forms that may be accepted as characteristic of the smooth phase of the avian, as well as of the bovine and human tubercle bacilli. Characteristic colonies of a smooth, human strain were shown in pl. 15, figs. 1 to 4, and were comparable with the smooth colonies of the bovine strain shown in figures 5 and 6. The rough colonies pictured in pl. 16, fig. 9 do not appear to be entirely characteristic, at least when compared with the rough form of avian strains, or with Christison's[17] photograph of her "variant III" of Brinkerhoff's bacillus, which seems characteristic except for the small central nipple. Continued marginal selections from such a colony would undoubtedly result, in a few transfers, in a perfect, flat rough. On the other hand, it is possible that the rough of the human type differs in morphology from the rough colonies of other acidfast species.

Making use of the culture phases mentioned above, Smithburn[174] found that, in human, bovine and avian strains, the smooth phase cultures were more virulent than the rough. On the other hand, there was one avian strain which retained the smooth form while at the same time losing its virulence. Smithburn's observation of smooth colonies in virulent cultures of human tubercle bacilli isolated from eight human sources other than sputum, seems to confirm the predicated pathogenic significance of smooth phase cultures of this organism in human infections as well as in experimental tests in animals. As to the place the smooth phase should occupy in avian and bovine infections, there can no longer be any doubt. On the other hand, the mucoid form of the tubercle bacillus has not yet been recognized with sufficient clearness to permit an opinion of its pathogenic significance.

In 1935 Birkhaug,[9] in a comprehensive study of variation in the tubercle group, investigated the comparative virulence of smooth and rough cultures of human, bovine and avian strains. He found that smooth avian cultures were highly virulent for fowl and rabbits but less so for guinea pigs, while the rough cultures were relatively avirulent for all animals. The smooth mammalian types were relatively virulent for rabbits and guinea pigs if cultures of the first or second passage were injected intraperitoneally or intravenously, but the virulence disappeared on further cultivation. The rough phase cultures of mammalian types were also reported as virulent—the human strains being pathogenic for guinea pigs, and the roughs of bovine origin for guinea pigs and rabbits. An exception was noted in the rough phase of B C G which was without virulence. While Birkhaug's smooth strain, "Ninni," appears characteristic, as indicated by his photographs (pl. 1, figs. 1 and 2), there may be some doubt as to the purity and stability of many of his rough phase cultures. He manifestly allowed considerable latitude for his interpretation of rough and smooth forms, and it is possible that some, designated as one phase or the other, were in reality intermediates—which, as I have often observed, are extremely troublesome when one is attempting to detect pure culture phases in a species as complex as the tubercle bacillus. It may be added that Birkhaug's monograph presents a considerable volume of other important aspects of dissociation in the tubercle group.

In summarizing these studies on the relation of colony form to virulence in the tubercle bacilli it appears probable that the smooth culture phase is as a rule the more virulent. Although the cultures studied from this viewpoint conform in a general way with the dissociative pattern observed in many other species, the situation obtaining in the acidfast group is unquestionably more complex, and the presence of intermediates as well as of other significant culture phases, must be accepted as a probability. Whether some of them will prove virulent is a problem for the future.

One special difficulty encountered in the study of dissociation in the acid-fast group is that in many of the species the long period of growth often required makes possible a sequence of changes even in the development of a single colony, so that a fair appraisal of culture phase may depend in some measure on the time when the colony observations are made, or injections of animals performed. I believe that, as a rule, observations should be made as soon as the morphology of the colonies can clearly be recognized. And to this it may be added that there is probably no other group in which the appropriate selection of a culture medium for adequate registration of colonies is so important, as clearly shown by Almeden[2] and by Cooper[217] in this laboratory. In virulence tests, moreover, it is just as important to ascertain what culture phase can be isolated from an animal after death as to know what culture phase is injected at the beginning. There is often no guarantee that they will be the same.

Turning now to another species, G. S. Wilson[204] in 1930 reported an unusually clearcut study on the "relationship between morphology, colonial appearance, agglutinability and virulence to mice of certain variants of Bacterium aertrycke." This work is of special interest because of the careful observation and characterization of the colony forms that were the basis of the study. With respect to virulence it was clear that the smooth forms were more active, but a point of special importance revealed by the author's photographs, was that the smoothest form was less virulent than some others in the smooth category that were not such perfect examples of smoothness, although Wilson did not comment especially on this circumstance. In other words the study seems to indicate that there existed differences in cultures within the smooth range. This raises the question whether there may not exist some S forms that are too "young" in the course of their development to possess full virulence, just as there are S forms that have advanced too far towards the rough phase to give characteristic results on animal inoculation. It is much more difficult to define what might be termed the "absolute S" than it is to recognize the "absolute R"; and this is because the "absolute S" lies at the midpoint of a series of gradual transformations, while the "absolute R" lies at the end of a transition process, beyond which the culture cannot pass without an abrupt change in all its morphological characteristics. In other words, although it is possible to recognize easily where the R phase terminates and the S begins, it is not commonly so easy to observe where the S terminates and the R begins. And this circumstance implies that our conception of what we term rough and smooth cultures should provide for a certain range of development in either case, rather than for narrowly fixed entities. Within these ranges, however, differences in both phases may appear.

Among the streptococci the relation between culture phase and virulence has been in an unsettled state for a considerable time, both smooth and rough forms having been presented as possessing the greater virulence. The situation has been such that Eagles[36] came to the conclusion that colony form could not be taken as an adequate indication of virulence. In addition, the mucoid phase has now been introduced further to complicate the situation.

Perhaps a part of the difficulty has been due to the lack of agreement regarding appropriate designations for the different forms of colony that have been described, and several of which cannot at present be brought within the common dissociative pattern of mucoid, smooth and rough.

In 1922 Cowan, working with strains of hemolytic streptococci, reported that she found the smooth phase cultures more virulent than the rough. In later years, however, this observation has not been entirely confirmed, and it seems possible that Cowan's "smooth" culture was actually a mucoid as Griffith[57] has recently suggested. The latter investigator, in 1927 reported three colony types. His type I (rough and flat) was probably equivalent to Todd's matt; his type II (soft, translucent) to Todd's glossy; and his type III to Todd's pseudoglossy—which, in turn, was probably the present mucoid. Griffith observed no difference in virulence between his coherent form (matt) and his watery (mucoid) types.

In 1927 Todd[185,186] reported the two colony types, matt and glossy, referred to above, but was unable to identify his colonies with those of Cowan, since Todd observed that the matt only possessed virulence. Subsequently Todd and Lancefield[187] reported the "pseudoglossy," which is probably identical with the mucoid. Also in 1928 Eagles,[36] after isolating what he termed rough and smooth strains, observed that the smooth phase of an erysipelas strain was more virulent, while the rough form of the Aronson streptococcus was more pathogenic for mice. Andrewes and Christie[5] in 1932 reported on earlier experiments in which they found the "rough" form about 200 times more virulent than the "smooth." In view of the recent studies of Dawson,[27,28,217] it seems possible that both Eagles and Andrewes and Christie were working with Todd's matt rather than with the actual rough phase; and that Eagles' "smooth" erysipelas strain may have been the mucoid.

Studying strains from erysipelas, scarlet fever and septic sore throat, Tunnicliff[190] observed that cultures from what she regarded as rough colonies were nonvirulent for mice while other colony types yielded virulent cultures. Among the latter were "typical rough, conical or convex colonies with regular edges," perhaps equivalent to Todd's matt type; also "large moist, convex, spreading, hemolytic colonies" that were presumably mucoids. Delves[29] working with one of the Dick strains observed small, smooth and conical colonies that she interpreted as smooths; also colonies which, from her photographs, might be regarded as perfect roughs, and which do not appear to be the counterpart of Todd's matt colonies. Delves reported that her smooth colony form was more virulent for mice. Cooley[19] also presented somewhat similar dissociative features in the strains studied by him. In the light of the reports of Todd,[186] Dawson[27] and of Ward and Lyons,[199] it may perhaps be questioned whether the colonies that Delves found virulent were actually the smooth in the sense of being equivalent to the "glossy" type of Todd. And it is also probable that the rough colony forms of Tunnicliff, Delves and Cooley were not identical with Todd's "matt," but were the actual streptoccus rough.

In the foregoing paragraphs there have been considered chiefly those contrasts in virulence which relate to the smooth and rough phase cultures. To what extent the picture may be modified by increased knowledge of the mucoid phase is still a question; and probably will be until a larger number of species have been studied with respect to this culture phase. In only a few species has this phase been submitted to comparative virulence tests, so that it is not known whether the situation obtaining in the pneumococcus, the hemolytic streptococcus and in Friedländer's bacillus with respect to virulence represents a special case of correlation between mucoid phase and virulence, or whether an analogous situation prevails in other species.

That an extension of this conception at least to the hemolytic streptococci is possible has been suggested by isolated observations ever since Davis[24] in 1912 reported his results on Str. epidemicus. One of the first, however, to attack the problem experimentally was Loewenthal[90,91,92] who published in 1932 important data bearing on the relation between culture phase, pathogenicity and antigenic structure. He observed four different culture forms, the N, M, O and R. Of these all were virulent for mice except the R (which was manifestly the rough), although the degree of virulence depended on the manner of inoculation. The M (mucoid) and N forms were highly hemolytic and toxigenic. The author stated that the M form was merely the mucoid form of the N. Since no photographs accompanied his report, one cannot interpret exactly the nature of the N and O forms. This study was later continued into certain immunological aspects of considerable interest, as will subsequently be shown.

The mucoid phase of the hemolytic streptococcus has also been dealt with in a preliminary way by Dawson and Olmstead,[28] who brought the culture phases of these organisms from human sources into closer harmony with the accepted dissociative scheme. They clearly identified the mucoid, the smooth and the rough phases. Regarding the distribution of virulence they stated that, although all of the cultures of high virulence observed by them were in the mucoid phase, all mucoid cultures were not virulent.

Ward and Lyons[199] in 1935 reported that, among four colony variants of hemolytic streptococci observed by them, only the "F" and the "M" (probably the mucoid) were isolated from the blood stream of patients; and that of these only the "M" possessed primary virulence for mice. Both were found to resist phagocytosis in human blood. Of the other culture forms observed one, the "attenuated M" was found in old cultures and was nonvirulent. The other, the "C variant," was conical in shape and was found most commonly in normal throats. It was nonvirulent and did not resist phagocytosis. This paper was continued by the authors into a study dealing with important immunological aspects of streptococcus infections as will be dealt with briefly on a later page of this review.

Among the greening streptococci the situation with respect to virulence is not clear because of conflicting reports and lack of definite study. Starting with Cowan's original observation that the smooth was more virulent, subsequent reports have shifted the responsibility from one phase to another, some presumably being neither smooth nor rough.

Tunnicliff and Woolsey[193] reported that 81% of cultures isolated from subacute bacterial endocarditis belonged to the viridans group; and that, of these, 84% "showed signs of roughness either morphologically or colonially or both." The authors believed that the rough element in the cultures might be an essential factor in the production of endocarditis. I am able to confirm the observation that the majority of such cultures in the primary isolation are rough. Such cultures, growing in a liquid medium have a tendency to colonize on surfaces.

Falk and associates[45] studied the virulence of smooth and rough cultures associated with epidemic influenza. Several tests appeared to indicate that the greater virulence was associated with the rough forms rather than with the smooth.

In contemplating the rather meager data on the rôle of the smooth phase greening streptococcus in infectious processes, one might ask whether there exists a mucoid phase; and, if so, what it is accomplishing in infectious processes. From the reports available it is apparent that this phase has not been commonly observed.

One lead into this problem has perhaps been supplied by Pilot[134] in 1934, and based on considerable first-hand experience with the bacteriology of the tonsils and nasopharynx during the influenza epidemic of 1917–18, as well as during the early post-epidemic period. Pilot's observations concerned in part the frequency-distribution of the so-called Mather coccus, also studied to some extent by Tunnicliff, after Mather's death. Pilot's suggestion was that this organism, which usually appears in the mucoid state and which is clearly distinguishable from Str. epidemicus, is really the mucoid phase of some variety of greening streptococcus, and the causative agent of epidemic influenza. Regardless of its possible relation to this disease, I believe that this suggestion is worthy of consideration in relation to the dissociative pattern of the green streptococcus. One reason for this view is the apparent conversion, in my laboratory, of a rough phase green streptococcus into the mucoid state by growth on a neopeptone blood medium. Its passage, rapidly through the smooth state is not, however, fully ruled out.

From the data presented in the foregoing paragraphs it becomes apparent that, among the streptococci, there exists a considerable diversity of colony form and that several different designations have been applied to colonies presumably in the same phase. There can, however, be no doubt about the existence of the mucoid phase or of the fact that Todd's "pseudoglossy" belongs in this type. It is also clear that the smooth phase is represented by the "glossy" of Todd, the S of Dawson and perhaps the C variant of Ward and Lyons. About the rough phase there still exists an uncertainty, bearing largely on the question whether the "matt" of Todd is to be regarded as the same as the R of Delves, Cooley, Loewenthal and Dawson, all of which latter forms lack virulence. It seems to me more appropriate for the present to separate these two forms until their individual characteristics can be more fully studied. In any case it seems clear that in the hemolytic streptococcus the attribute of virulence is divided between two different phases, the matt and the mucoid, of which the latter only carries a high degree of pathogenicity for mice. The degree of virulence possessed by the matt form in human tissues is still a question. One situation that still complicates the matter is that there are both matt and mucoid cultures that appear to lack virulence for mice, as shown by Todd and Dawson, respectively.

In concluding this section on the relation of colony form to virulence it may be said that, although the mucoid nature of cultures that possess maximum virulence for mice has now been demonstrated for the pneumococcus, hemolytic streptococcus and the Friedländer bacillus, the rôle in infection played by this phase in other pathogenic bacterial species is not yet commonly recognized. But the fact that there exist so many reports of capsulated organisms appearing in cultures fresh from infected tissues justifies the suggestion that the organisms at work in the body may be frequently in the mucoid phase; and that this phase might be demonstrated more often in primary cultures if an appropriate medium were employed. It seems clear that the use of neopeptone by Dawson and by Ward and Lyons facilitated the appearance, and perhaps maintenance, of the mucoid forms of streptococcus. On the other hand, both Pilot and Jennings found that ascitic fluid was of considerable value for purposes of smooth to mucoid transformation; while Oppenheim obtained good mucoid development on both the Bieling plate (blood-water agar) and chocolate agar. Although the chief colonial forms of streptococcus are clearcut entities when they appear on a favorable medium, it seems probable that some of the unusual colony manifestations are due to the influence of the substratum, as also in the case of the tubercle bacillus.[2]

Toxigenicity.—The earlier conclusion that smooth phase cultures were more active biochemically than the rough, and that their toxigenic power was greater, has received additional confirmation in recent times, although there are occasional exceptions.

Thus Yü[211a] demonstrated that smooth phase cultures of the diphtheria bacillus produced more toxin than did the rough; also the important point that purely antitoxic serums possessed no ability to cause a transformation from the smooth to the rough phase. These observations were confirmed by Morton[115] and myself, also by Neisser.[117] Orr and Reed[148] observed that the power of elaborating hemotoxin by the smooth phase of the Welch bacillus was ten to twenty times that possessed by the rough. McGaughey[103] stated that his variant II produced both toxin and hemolysin in greater amount than did his variant I, or the "normal" form. It appears, however, that in this case both I and II were well within the rough range as judged by the photographs, while the "normal" form was the smooth.

While the majority have reported a higher toxin-producing power in smooth strains of species of the enteric group, Maltaner[98] observed no difference in the degree of toxicity for rabbits of R and S cultures of the typhoid bacillus when the intravenous method of injection was used. Intraperitoneal injection into guinea pigs, however, showed the smooth form to be more toxic.

In the streptococci Todd[186] in 1928 showed the "matt" and "glossy" forms to have about the same degree of toxigenicity; moreover, that there was no correlation between virulence and the factor for toxicity. Eagles[36] indicated no difference in degree of hemolysis between his rough and smooth forms. Loewenthal[91] reported that both his M and "N" phase were more toxic and hemolytic than either his "O" or R. Oppenheim,[126] on the contrary, stated that the mucoid was more virulent, but less hemolytic, than the alternate phase. My own observations have been that, as a culture develops more virulence by acquisition of the mucoid state, hemolysis is greatly reduced, and sometimes lost. Oppenheim's mucoids gained hemolytic power on transformation to the smooth phase.

Wheeler,[202] studying three epidemics of septic sore throat, in New York State in 1930, observed that in only one was a typical Str. epidemicus isolated; and that this one was a toxigenic strain. Of the cultures isolated from the other two epidemics, one gave a strain of low virulence in which capsules were absent while the other showed no capsules and generated no toxin. Of these two strains, the former, when passed through mice, transformed into a culture of the epidemicus type (mucoid). While the author failed to present any detailed observations on the colony forms, which would at present be of great value, it appears probable that she was dealing with the mucoid and smooth phases of the hemolytic streptococcus; and that toxigenicity was limited to the mucoid phase. This relation has also been observed by Pilot.[134]

In 1931 Leslie and Gardiner[86] who were particularly interested in the serological types of B. pertussis and their relation to the known culture phases, observed that, of the four recognized serological types, I and II were smooth forms while III and IV were rough. The smooth I was more toxic and the authors concluded from tests on guinea pigs that vaccines made from this phase showed the greatest value in the production of active immunity. This work is of special interest because it demonstrated the fallacy of "serological types" reported for this and that bacterial species, when these "types" are founded, not upon different type-specific carbohydrates or protein fractions, as in the pneumococcus and the streptococcus serological groupings,[85,57] but on a diversity of culture phases. Some years earlier, Atkin in England had revealed a similar situation with reference to the "meningococcus types."

Conclusion.—This section relating to virulence and toxigenicity can be summarized by the statement that, in the majority of cases carefully studied, it still appears that (among the Bactericeae) rough phase cultures as a rule lack both virulence and toxigenicity, although certain exceptions are to be noted—particularly among the spore-forming aerobes and anaerobes. In species in which only the smooth and rough phases are clearly recognized, it is the former to which virulence is most common attached. On the other hand, among those species in which the mucoid is more common, and perhaps in some others as well, this phase may possess maximum virulence—although nonvirulent mucoid forms are known. Among the streptococci, however, the "matt" colony type is more virulent than the smooth. Regardless of the situation with respect to the pneumococcus, the streptococcus and the bacillus of Friedländer, it is not yet established that, among species in general, the mucoid always carries higher virulence. The colony phase associated with toxigenicity cannot yet be established; although it is apparently not the rough. There is no biological reason for assuming that a culture possessing high powers of invasiveness must, at the same time, be more toxigenic or hemolytic. At present it does not appear that hemolysis is correlated with any special phase; indeed, there is some evidence of a negative correlation between hemolysis and mucoid phase. With respect to its immunological possibilities, this phase offers many opportunities for further study, and this holds not only for those species in which it is common, but in others as well. A great amount of work remains, however, to be accomplished before it will be known what biochemical attributes characterize the chief phases; and it is possible that correlations that are established in one species will not be duplicated exactly in another.

DISSOCIATION IN RELATION TO EPIDEMIC WAVES

It is perhaps natural that the observations of recent years on the more exact nature of the conditions that underlie virulence and toxigenicity should stimulate renewed speculation on that inscrutable problem that has for so long occupied the minds of bacteriologists—namely, the cause of epidemic waves. These speculations, starting in early times with the influence of the constellations and subsequently coming down to the weather and the distribution of "susceptible material" in the general population, have now to a degree focussed on the ups and downs in the life of the bacteria. Notwithstanding the large amount of work that has been done in recent years in the field of experimental epidemiology, it has not brought us very far on the way toward a solution. Neither has the meager knowledge thus far gained regarding microbic dissociation supplied any fact definitely interpretable as bearing on the epidemic problem. Despite this situation, there have been some grounds for examining the dissociation aspect, and the views of a few bacteriologists interested in this problem may be cited briefly.

Jordan,[76] partly perhaps on the strength of his general appreciation of certain elements of truth in the dissociation concept, and in part from his own valuable observations on the changes in phase and in virulence among paratyphoid bacteria, was among the first to abandon to a degree the earlier notions regarding the influence of the amount of "susceptible material" distributed through the general population. He considered it more probable that, through dissociation or through some other manner of variation, new and more virulent strains of organism came into being. Jordan's conclusions were also based upon his extensive observations on the influenza epidemic of 1917–18.

Falk and associates[45] were subsequently led to give even more direct expression to the view that, with respect to influenza, microbic dissociation and the accompanying transformations in virulence played a significant part in the determination of epidemic waves. These views were directed particularly toward the greening streptococcus which Falk regarded as the causative agent in influenza. The observations made by Kun and Fenyvessy[84] already referred to, on the dissociation of the influenza bacillus, including the very slow recovery of virulence by their rough phase cultures, gave basis to their suggestion that similar gains and losses of virulence might play a part in the epidemiology of influenza.

In 1932 Zinsser[214] gave consideration to the same problem. He accepted the premise that, accompanying the in vitro and in vivo transformations of bacteria, changes can take place both toward and away from virulence and pathogenicity, and that it therefore "remains only to define the conditions under which, and in which direction, these changes occur." Although admitting the temptation, in efforts to explain epidemics, to include changes in virulence, he felt that experimental epidemiology "gives no basis for assuming that enhancement of virulence is a factor of importance in the rise of an epidemic wave; or that the increase of avirulent over virulent individuals organisms . . . has anything to do with subsidence." Indeed he observed the present tendency among epidemiologists to exclude the facts of microbic dissociation. Despite this, he was inclined to agree with Stallybrass that, in epidemiological experiments, the imposed conditions might not reproduce community conditions. He believed further that there might be some danger of underestimating the influence of natural variations in virulence of the inciting organism. The reasons for this view were expanded by citation of instances from epidemics in cholera, plague, smallpox, poliomyelitis, diphtheria and other infections. Zinsser also proceeded to construct, mathematically, a picture of what might happen in relation to the spread of an epidemic in a hypothetical case and under hypothetical conditions. But the present writer does not understand just what conclusions should be drawn from this display.

As the expression of the view of an experimental epidemiologist who, as Zinsser remarked, "seems to have done everything that is possible to do with a mouse population to throw light on this question," Webster concluded that microbic dissociation, with its alternations in virulent phase, has slight if any influence on the rise and fall of epidemics. Webster's most important contributions have appeared in the Journal of Experimental Medicine.

In 1932 I[68] discussed briefly the possible relation of microbic dissociation to epidemics and concluded with the view that "until more information can be obtained it may be helpful to consider two possibilities: (1) that the epidemic nature of certain communicable diseases may be due to the slow, evolutionary transition in culture phase occurring in the infective agent present in many of the population; (2) that the morbidity rates in some epidemics may be influenced by the extent of dissemination of the organisms of nonvirulent phase during previous inter-epidemic periods."

To this it may now be added, as against the possible view that dissociative changes occur too rapidly to be in harmony with observed facts relating to epidemic waves, that a renewal of virulence in culture phases that have become thoroughly attenuated occurs much more slowly than does the loss of virulence in virulent phases. This is naturally in harmony with the further observation that cultures once having attained the avirulent (rough) phase are much more difficult to move again toward the virulent state; and that, in some instances, attaining a location in living tissue is the only stimulus that can start the reversionary process. When the ultimate transformation does occur, however, it is likely to be rather sudden.

Summarizing this aspect of the dissociation problem, it may be said that, as yet, there exists no definite experimental evidence of the influence of dissociative behavior on the rise or decline of epidemics. At the same time, there are gradually accumulating certain observations which suggest that it is in this field of observation and experiment that one should look for the final solution of the epidemic problem.

DISSOCIATION IN RELATION TO PROBLEMS IN IMMUNOLOGY AND THE COURSE OF THE INFECTIVE PROCESS

If the culture phase possesses special significance with reference to the degree to which infectious processes may be instituted and maintained, as appears to have been clearly shown, it might be assumed that the proper choice of a culture phase to be employed in a bacterial vaccine, or used for the production of an immune serum, might be equally important.[68] Although some additional study has been devoted to this subject in recent years it has not been carried far except in the case of the typhoid bacillus and the hemolytic streptococcus; and all conclusions, so far as practical advantages are concerned, must for the present be held in abeyance.

Prophylactic vaccination.—In this field the results of recent studies have produced some confusion, although the significance of the dissociative concept for studies of this sort has been amply indicated. In harmony with earlier views that, as a rule, the rough phase cultures were neither pathogenic nor of value in immunizing procedures, Grinnell[58] reported that the blood of human beings vaccinated with the rough Rawlings strain of the typhoid bacillus revealed much less germicidal power than the blood of individuals immunized with freshly isolated smooth strains. In 1932 the same author[59] demonstrated that the Rawlings rough was much less efficient than smooth strains as a protective antigen. The experiments supporting these conclusions involved virulence tests in immunized mice, and also estimations of the bactericidal effects of fresh guinea pig blood. The results of the study were such as to indicate to the author the desirability of substituting smooth virulent cultures for the very old Rawlings strain in typhoid vaccination. Grinnell's results were in harmony with many earlier reports demonstrating the absence in immunizing value of rough phase variants in other bacterial species.

In contrast with these results are those presented in 1934 by Maltaner[98] on prophylactic vaccination against the typhoid bacillus in animals. He reported that cultures of both rough and smooth variants, administered intravenously into rabbits, gave subsequent protection against approximately 14 times the minimum invasive dose of living, smooth bacilli; but "the rough vaccines protected distinctly better than the smooth." The carrier state in rabbits was produced exclusively by the smooth culture, while toxicity in rabbits was "distributed about equally in both." At the same time it was reported that the smooth cultures given to guinea pigs intraperitoneally were more toxic than the rough. Maltaner concluded that his results were "in conformity with the practical experience of the past thirty years in the preventive inoculation of human beings against typhoid fever."

The manifest discordance between the reports of Grinnell and Maltaner is difficult to explain. Although the latter's description of his rough phase cultures suggests that they were characteristic, his report was not accompanied by photographs of either the rough or smooth forms, and I have often been led to doubt the phase purity of the so-called "motile rough" strains. Such a discrepancy in colony diagnosis is therefore possible but not probable. Moreover, Maltaner's serological tests seem to have given evidence of distinctive rough variants. At present about the only explanation of the situation appears to lie in an observation made by Felix dealing with his "Vi" antigen of the typhoid bacillus.

In 1934 Felix and Pitt,[49] continuing their studies on the antigenic structure of the typhoid bacillus, and especially with reference to the O and H components and their respective antibodies, showed that virulence was due specifically to the ability of the organisms to resist the influence of the O type antibody. While they found agglutinable strains to be of low virulence, inagglutinable strains were of high virulence. Since both strains were related in the quality of smoothness, and did not differ in the content of O receptors, it became apparent that the mere presence of smooth O antigen did not, as formerly believed, determine by itself high virulence; and that some still unrecognized factor was required to render the O antigen resistant to the O antibody.

Omitting the details of their experiments, it may be said that this missing factor was discovered by Felix and Pitt in the guise of a special antigenic factor determining virulence — namely, the "Vi" or virulence antigen. The Vi antibody, purified by adsorption technic, was found to have the ability of agglutinating the virulent strains. It was also found to possess upon the injection of animals a strong protective influence in both an active and passive sense, while the O antibody exerted distinctly bactericidal and opsonizing effects, and neutralized the typhoid endotoxin. The Vi antibody did not possess this ability. Further observations on the Vi antigen were subsequently reported by Felix, Bhatnagar and Pitt.[51]

But the point that is of special interest in relation to the previously mentioned experiments of Grinnell and of Maltaner was reported by Felix and Pitt[50] in 1935 and concerns the presence of the Vi antigenic component in the various culture phases of B. typhosus. Here they stated that the integrity of this component was not necessarily lost during the transformation from the smooth to the rough phase, although the O antigen, or its specific polysaccharide component, might disappear in the process. They could not state which occurred more frequently—the loss of Vi or the loss of O.

Returning now to Grinnell and Maltaner, and the comparative value of smooth and rough cultures for immunization, the question naturally arises: Did the R phase culture of Maltaner contain the Vi antigen, although this appeared to be lacking in Grinnell's rough Rawlings strain?

To the above it may be added that, although Felix and his associates have added greatly to our knowledge of the antigenic structure of the typhoid species they have not been concerned in any detailed manner with the problem of culture phases, and have not complemented their findings by clear descriptions or photographs of their colony forms. For this reason there exists, as in many another instance, an unfortunate detachment between these aspects—both of which are important components of the larger problem. Moreover, it seems almost certain that a mucoid phase of the typhoid bacillus must exist, although not yet definitely recognized as such; and, until this is discovered and studied with reference to its chief biological attributes, including the serological and antigenic, the full antigenic nature of the typhoid species will not be known.

Immune serums and specific opsonification.—Closely related to the problem of effective prophylactic vaccination with selected phase antigens is that of the employment of such antigens for the production of immune serums. This is, in turn, in many instances at least, related to the question of immune opsonins and the degree of resistance to phagocytosis manifested by organisms of different phase; for it is becoming increasingly apparent that, aside from its antitoxic features, the protective influence in an immune serum depends in considerable degree on opsonizing effects.[188a] Although all the important recent studies in this field cannot be reviewed here in detail, the investigations of Tunnicliff, Hare, Loewenthal, Ward and Lyons and of Mellon, all concerned with immunological aspects of antistreptococcus serums, are of special interest and will be presented briefly. In so doing it will be impossible to introduce the fascinating picture produced by the work of Lancefield, Todd, Griffith, Andrewes and Christie, and others relating to serological type and group specificity in the streptococci. With reference to their investigations it may only be said here that, through the new recognition of certain antigenic components, protein or carbohydrate, there have been effected grounds for serological differentiation that, for the first time, are beginning to put some degree of order into the large streptococcus group.

Although Hare[70] did not concern himself appreciably with streptococcus cultures of known phase, he did distinguish the virulent and nonvirulent forms and reported on the differential phagocytosis of such cultures, with special reference to strains derived from the puerperium. The slighter degree of phagocytosis of the more virulent cultures was clearly shown. He also studied alterations in the bactericidal power of the blood occurring during the course of streptococcus infections, and concluded that in patients with localized infections (uterus or immediate neighborhood), any observed increase in the bactericidal power of the blood, if it occurs at all, appears too late to explain the absence of general invasion; while, on the other hand, "in patients with generalized infections who recover, the bactericidal power of the blood is much greater than normal." This property was attributed to both direct serum influence and to phagocytosis. Such host factors as those recently emphasized by Locke,[89b,217] were not taken into consideration, although it is probably true that these possess greater significance in the defense against intending infections than in the elimination of infections once established. It may be added to the above that Hare's virulent strains were probably in the mucoid phase. Seastone[163] has also correlated diffuse growth in broth, resistance to phagocytosis, presence of capsules and virulence.

The studies of Loewenthal[91,92] were characterized by the employment of definite culture phases of hemolytic streptococci in prophylactic vaccination and prophylactic serum treatment in mice. Employing culture phases designated M, N, O and R, he demonstrated the high degree of active immunity that could be produced with N and O vaccines. The N form was type-specific while the O form was only group-specific. The immune serum derived from the N form protected mice, but only against the homologous culture. The serum derived from the "capsulated O form" was highly protective against the homologous form when tested by either prophylactic or simultaneous administration. None of the serums possessed apparent protective value when administered after the infecting dose.

Although, as in the use of pneumococcus serum, it is realized that protection in mice and in human beings are two quite different things, and although there exist discrepancies between mouse virulence and human virulence,[199] the work of Loewenthal has considerable significance in that it probably represents the first clearly reported attempt to relate definite culture phases of the hemolytic streptococcus to serum protection. Perhaps the chief disadvantage for others in Loewenthal's presentation is the lack of adequate description and photographic records of the culture phases employed in the experiments. This situation is another example of the need of reaching some agreement regarding appropriate designations for the chief, and probably other, culture phases of the streptococcus.

The studies of Ward and Lyons[199,95] concerning dissociation of hemolytic streptococcus in relation to the immunology of streptococcus infections have presented additional points of interest. Their experiments offered a natural sequel to certain earlier observations by Hare and by Seastone, but possess the additional advantage of taking cognizance of some of the chief phases present in the cultures with which they worked; also the advantage of extending their inquiry to the study of immune serums whose opsonizing value for certain strains of steptococcus was previously recognized, particularly through the earlier studies of Tur nicliff.[190 192a] She found the opsonic method the most satisfactory one for differentiating certain dissociants of hemolytic streptococci, including strains from both erysipelas and scarlet fever. As early as 1920 Tunnicliff[188a] had shown that specific mouse protection by antistreptococcus serum varied with its content of specific opsonin. Ward and Lyons again demonstrated that at least one important factor in the mechanism of serum treatment in streptococcus infections is bound up with the specific opsonins, and that these must be related to organisms in the mucoid phase. With respect to the recognition of virulence in streptococcus cultures these writers abandoned the mouse test in favor of phagocytic tests as an indication of virulence for man; and a special technic for conducting such tests was presented.

One important aspect of the immunology of streptococcus infections touched upon briefly by Lyons[94] and by Lyons and Ward[95] dealt with the characteristics of an antibacterial as opposed to an antitoxic serum, and reviving an issue first suggested by the work of Yü[211a] for the diphtheria bacillus, confirmed by Morton[115] and myself some years ago. Yü was able to demonstrate that a purely antitoxic diphtherial serum possessed no ability to transform the smooth, toxin-producing phase of this organism into the rough, nontoxigenic form; while an antibacterial serum was able to accomplish this. In a somewhat similar manner, Lyons and Ward demonstrated that at least certain commercial antistreptococcal serums, while capable of blanching the rash in streptococcus infections, showed no opsonizing effect in vitro. This appeared to be possible explanation of the failure of some scarlet fever serums to interrupt the invasive course of the disease process while at the same time being of value in neutralizing the toxins in vivo. The studies of Yü and of Ward and Lyons thus concern a practical issue in serum therapy which merits additional investigation.

In addition to the above studies on the immunology of streptococcus infections, and its relation to dissociation, certain still unpublished observations and experiments of Mellon[106,217] may be referred to. His earlier observations[81a] on the relationship between streptococci and diphtheroids suggested to him a new approach to the treatment of streptococcus infections—namely the possible preparation of a serum from some special phase of the organism that might effect an in vivo dissociation of the virulent form into less virulent phases, such as diphtheroids and rough forms. In 1925 two anomalous variants were obtained and from them immune serums were prepared in horses. The serum which has been employed by Mellon most generally for treatment of human cases of streptococcus infection has been shown to contain opsonins for a number of strains of streptococcus isolated from these cases; but a greater number of freshly isolated strains in virulent phase must be tested before its range of influence can be established. The serum has now been administered in 120 cases of severe streptococcus infection, such as cellulitis and erysipelas, with a resulting mortality rate of 12%.

Mellon has not reached a conclusion as to the extent to which the antibodies in this serum function to stimulate the mechanism of recovery, since he has observed that dissociations may be accelerated by certain general influences as well as by specific immune serums. It has, however, become apparent that in repeated instances, in serum treated cases, in vivo dissociations to smooth phase cultures or to diphtheroids has been observed to accompany or follow the clinical crises. But only by the observation of a larger number of cases can it be concluded whether these instances are representative. Mellon believes that some indications point to the existence of some sort of "tissue immunity," since he has observed considerable fluctuation of the calcium-phosphorous ratios in the blood of patients, these frequently being considerably lower during the active stages of the infection.

Dissociation in relation to the infectious process.—Closely related to the problem of immune serum influence, either in the culture tube or in the body of the infected animal or patient, is the matter of response on the part of the organisms to the serum environment. This question concerns a possibility increasingly present in the minds of immunologists of controlling infections, not by an attempted destruction of the invading organisms directly, but by forcing them out of the virulent phase into one in which they are less invasive, or in which they become susceptible to phagocytosis—which, in either case, plays the final rôle. While few attempts to study this problem as a separate issue have been made, aside from the reports previously cited, a few observations that bear closely upon it may be presented.

In 1927 Wardsworth and Sickles[200] reported that, when the pneumococcus (mucoid phase) multiplies in the tissues of immunized horses, it loses its virulence, its power of forming capsules, its resistance to phagocytosis, and its type-specificity. In other words, it becomes transformed from the mucoid to the smooth phase. The authors could not conclude whether these changes were to be attributed to the influence of the immune horse serum in vivo, or to an "adaptive process" on the part of the organisms to an adverse environment. Today, however, both they and others would probably regard the answer as clear.

Shibley and Haelscher[168] in 1932 observed rough forms (O.D.) of pneumococcus in 13 of 16 successive colony studies of cultures taken directly from the lungs in cases of lobar pneumonia; and they regarded it possible that the S to R (O.D.) transformation might play a part in recovery.* Dawson[27] and Tunnicliff[189] have also observed changes in the phase of hemolytic streptococcus isolated from patients at different times in the course of the infection and in convalescence. I have noted transformations in cultures from patients under treatment with Mellon's[106] streptococcus immune serum at this hospital. There is not, however, much unanimity at present regarding the nature and direction of this transformation, except that Mellon and Koch[81a] have reported the appearance of the diphtheroid form.

Reimann[149] showed that when pneumococci in mucoid phase grew in agar foci in guinea pigs, there occurred some degree of transition to smooth. Paul[129] produced dermal lesions in rabbits, mice and guinea pigs by the simultaneous injection of rough phase (O.D.) and certain irritating substances. There occurred dissociations in which the culture might revert to the original form, or become "degraded" to forms that were even "rougher" than the primary culture. The former reaction was seen only in mice, while the latter occurred more commonly in rabbits. It seems possible that, in these studies, Paul had in hand certain colonies that would today be designated as the rough phase pneumococcus. The author stressed the point that the course of phase transition depended largely on the kind of environment surrounding the organisms.

Studies in this field possess considerable value because they assist in understanding the significance of the in vivo dissociations of bacteria that have entered the body under natural conditions of infection, whether this concerns transformations toward greater virulence or transformations toward a more attenuated state. For the results of these studies should eventually indicate whether there yet remains to be defined a new aim and method for combating infectious diseases—namely, by controlling the culture phase of the infecting agent.

BIOLOGICAL SIGNIFICANCE OF MICROBE DISSOCIATION
(DISSOCIATIVE VARIATION)

While the majority of bacteriologists have long been content to report objectively what they could discover, regarding the superficial aspects of bacterial variability, a few in each generation have possessed the urge to peer

---

* It appears probable that this perhaps more or less continuous transformation from M to S in the lung tissues of pneumonia patients explains the common observation that, in the typing procedure by the Neufeld technic, there are usually encountered many pneumococci that do not present the characteristic capsule swelling in the presence of the homologous type

below the surface of appearances and to inquire into the biological significance of the observed changes. After all, why is the smooth or the rough phase culture what it is? Why does the mucoid transform to the smooth before passing on to the rough? Why is one culture phase virulent and another not? Moreover, is all variability observed among bacteria of the same order? And, if there exist different orders, what is their cause? Is microbic dissociation to be regarded as synomous with bacterial variability? And, if not, what are their respective definitions and limitations? These and many other similar questions come insistently into the minds of those who would like to see bacteriology take its place more securely as a real science, instead of devoting all of its energies in the pursuit of attempted applications. While there appears to be no immediate prospect that the great problem of variability can soon be solved, there do exist reasons for believing that it can be simplified.

Possible variation categories.—So far as interpretations of bacterial variability have been attempted, they have fallen into one or another of five main groups. The variant has been regarded as an example of: (1) fortuitous variation in the Darwinian sense; (2) mutation in the sense of de Vries; (3) orthogenetic variation; (4) effect of environmental factors (impressed variation), necessarily taking on the aspects of Lamarckianism; and (5) progressive ontogenetic transformations, commonly projected into a conception of life cycles (cyclogeny).

None of these conceptions has lacked disciples. Probably the majority of bacteriologists have supported the theory of fluctuating variations and the theory of environmental influence, calling to their assistance in times of need the mutation hypothesis—but only when the new variant appeared so suddenly, or departed so widely from the norm of the "species," that there appeared to be no other choice. Orthogenesis has seldom been invoked, chiefly perhaps because bacteriologists have seldom examined with sufficient care the trends of culture development in different species. Those who have been drawn to this explanation, however, have clearly indicated by so doing that they have been sufficiently alert to see one of the most significant facts of dissociative variation; but they have not seen far enough. As for the theory of progressive ontogenic (or cyclogenic) transformations, the supporters have been few; but as a rule they have represented a group of workers whose main interest has, for many years, centered in the variation problem. And because of this, they are, perhaps, more than others entitled to an opinion.

Regarding the influence of fluctuating variations, since they occur in all living things, no one will dispute their probable occurrence among bacteria; but it is apparent that they are not identical with the sort of variant that is the center of interest in studies on dissociative behavior. It was the directive trends in the variational process in mycoides (and perhaps recognition of the parallel trends in many other bacterial species) that led Lewis to see in orthogenesis a possible explanation of bacterial variability. While this interpretation may be regarded as, in a sense, constructive, since it took cognizance of one important feature of variability, the failure of Lewis[87] to translate his observed facts into ontogeny, rather than into orthogenesis, was manifestly due to his failure to observe "a well marked chain of events leading back in reverse order to the original basic form." Therefore, confronted by a choice between ontogenesis and orthogenesis, he took the latter.

While the influence of environmental factors, physical or chemical, in modifying the form of culture growth and biochemical reactions, and to some extent in altering the morphology of bacterial cells, has never been doubted by bacteriologists, there has of late been an increased tendency to regard environment as responsible for some of the most important phenomena of variability, including the chief forms observed in dissociative variation. That so-called "impressed variations"—to use a term long ago introduced by Winslow —frequently occur, no one will deny; but that they play any part in the larger problems of variability is highly improbable. Indeed, the frequent confusion between orders of variability phenomena that are significant and those that are not, serves considerably to delay the acquisition of useful knowledge in this field of study.

Among recent publications in which special insistence is made upon the influence of environmental factors, operative over relatively brief periods, as the cause of variation, are those of Rettger and Gillespie[153,154] dealing with variations in B. megatherium. In this study emphasis was placed on variations in cell form, while the relation of cell form to colony type and culture phase was given no consideration. The authors pointed out the influence on cell form of several environmental conditions and believed that they largely embraced factors "unfavorable to continued normal growth." They stated: "In other words variation is possible only when favorable and unfavorable influences are so balanced as to permit slow growth in the face of untoward circumstances." They desired to "stress again the importance of subjecting cultural material to rigid and continued microscopic examination in life cycle studies."

Perhaps the chief criticism of the work mentioned above is that the conditions under which the authors studied the variability of megatherium were scarcely adequate to permit the manifestation of a cyclical development, or even of dissociative trends, had they really been present in the cultures. No one doubts that changes in cell morphology occur in cultures submitted for brief or for long periods of time to this or that new environment, but such changes are not necessarily characteristic of, nor identical with, those transformations that are accompanied by colony modifications, and are therefore regarded as significant by those favoring a cyclic or ontogenetic hypothesis. Indeed such a sort of experimentation may be questioned as to having any significant bearing upon the fundamental study of bacterial variability— other than to show that changes in environment can modify cell form; and that is scarcely a new or important observation.

One of the widely discussed problems in bacteriology today, and one intimately associated with variability, is the existence of life cycles. Those opposed to them, commonly speak of them as "complicated life cycles." To postulate a cycle, however, does not necessarily make it a "complicated" one; in fact some might be regarded as being fairly simple. Many bacteriologists have now abandoned monomorphic conceptions to such a degree as to accept the chief phase variants, and to adopt them as a means of elucidating many another problem in bacteriology, serology and immunology, without feeling at the same time the necessity of reaching a conclusion regarding the part such variants may play in the life cycles of bacteria. And that is a position that can scarcely be criticized, since this is rather a problem by itself and far removed from the sort of interest that most bacteriologists have in bacteriology; moreover, it is a problem that requires some reading and some thinking and considerable experimental endeavor, or perchance all three, before any opinion is worth very much.

But there is another group of bacteriologists, accepting equally the existence of phase variants as orderly and respectable members of bacterial society, whose interest happens to center more upon their significance than on what can be got out of them in the pressing affairs of medicine, agriculture or industry. And it is from this group that there arise, year after year, sometimes vague, sometimes clearcut, intimations of a cyclical trend in bacterial development, hidden in the variation processes of bacteria.

And finally, there exists a third group, given over to defending bacteriology against the malign influence of over-zealous pleomorphists with their threats of life cycles—and even occasional hints of the more repugnant phenomena of sexual reproduction among bacteria. It is sometimes even feared that the entire scientific edifice of bacteriology will topple unless life cycles and filterable forms are held in check: "If bacteria must pass through various life cycle phases, and if they manifest themselves in viable forms that are filterable, our conception of bacteriology will require thorough revision, and with it the present conception of pathology and medicine." [But] "If what are considered by many as life cycle phases of bacteria are merely variants which have no cycle significance and which are due to environmental influences, or certain inherent tendencies to vary from or around a norm, and if the so-called filtrable forms of bacteria do not exist as such—except perhaps occasionally as isolated cells or as broken bits of bacterial cells, the threatened cataclysm in systematic bacteriology may be indefinitely deferred."[153]

In commenting on the quotation given above it may be ventured that many bacteriologists will not regard the situation so threatening as to demand the cessation of their possibly constructive work on the problems of life cycles and the filtrable forms of bacteria. The science of bacteriology has possessed sufficient fortitude to withstand, thus far, the injection of what were, not long ago, distinctly "radical notions" regarding the existence and importance of microbic dissociation and the chief culture phases. And it may be anticipated that, even if "life cycles" and "filtrable forms" eventually need to be added, they will be borne with courage and equanimity by a science that is, I believe, just entering into its greatest period of self-constructive activity.

Life cycles of bacteria.—This constitutes a problem so difficult of approach that advance is slow, and probably will be for many years. Regardless of the frequent free expressions of opinion in contributions whose experimental data scarcely justify an opinion for or against the hypothesis, the authentic and relevant data are accumulated only slowly, and by those whose main interest is, and has been for many years, the problem of bacterial variation. In the entire history of bacteriology not many names appear in this group—only perhaps Frohmann, Almquist, Hort, Löhnis, Mellon and Enderlein, to which group should probably now be added the name of Cunningham whose study of the variability of B. saccharobutyricus is perhaps the outstanding contribution to this subject in recent years.

Cunningham's papers[22] appeared during 1930–31 and were based on the study of 20 strains. Among the variants observed were many organisms that had previously received other specific names, such as M. candicans, M. aurantiacus, M. roseus, B. circulans, Pectinobacter amylophylum and B. brevis. Among the cell mechanisms through which the transformations were apparently accomplished were the formation and germination of endospores, exospores and microcysts; also the formation of gonidangia, gonidia and regenerative bodies. His observations also mentioned the formation of, and regeneration from, symplasms; also conjunction followed by the development of regenerative bodies at points of contact. Certain features of this work with special reference to exospores as a transition mechanism have been confirmed by Mellon and Bambas[217] at this institute in still unpublished work. For the generation of his variants Cunningham did not employ special mediums nor unusual exciting substances or conditions of growth. The author found it impossible to arrive at a conclusion as to the nature of the factors controlling the generation of these variants. This circumstance furnishes additional evidence against the "injury hypothesis," too often called upon to explain cell and culture transformations.

The relation of his variants to culture phase was not touched on by this author, nor perhaps could have been, considering the magnitude of his problem. But it seems clear that many, and probably the majority of his variants transcend the range of dissociative variation as commonly recognized. I believe that this may be taken to mean that dissociative variation is only one aspect, and probably the simplest one, of the greater variation problem. The manner in which the phenomena of microbic dissociation, as now unfolding, may be related to the broader problem will be considered briefly in later pages of this review.

Writing about 1929 in his splendid chapter on "variation" in the English System of Bacteriology, Arkwright[7] expressed the view that it is still a question whether S and R forms represent stages in a cycle. He felt that most of the present evidence for cyclogeny in the sense of Enderlein "rests largely on doubtful theoretical grounds and forced analogies." He believed that the most acceptable view at present was that bacteria are "simple forms which can comparatively easily be modified by environment and can pass on such changes to their offspring." It is thus clear that Lamarckian imputations did not bother Arkwright. In a later publication he[6] further suggested the possibility that some changes that appear cyclical may be conditioned by environment; also that some "temporary variations, induced by environment, may sometimes simulate real cyclical changes." Regardless, however, of the extent to which bacteria might be "moulded by their environment," Arkwright also saw evidence for the existence of "a strong hereditary force which maintains the specific identity. . . ."

Van Loghem[195] did not favor the cyclical hypothesis but observed that the klone played a part in variability,—this representing the asexual descendents of one bacterium. This progeny inherit not only the genotypic characters, but continue in a somatic manner the individuality of the parent cell. It appears to me that this view of variation extends too deeply into matters that are not very well understood by the geneticists themselves, to be of much help in elucidating the problem for the bacteriologist. The botanists are in a position to do better, though they have not, as yet.

Mellon, Richardson and Fisher[113] reported very briefly certain biological sequences in the avian tubercle bacillus which they regarded as a life cycle. They presented the following scheme: The smooth phase culture (stage 1) gave rise to filtrable forms (stage 2). These in turn generated into a diplococcus (stage 3) followed by a large tetrad (stage 4). This subsequently gave rise to a stabilized diphtheroid (stage 5) which later produced acidfast gonidia (stage 6). These in turn transformed into a culture of the rough phase (stage 7), subsequently converted to the smooth (stage 1). Stages 1, 6 and 7 were acidfast, while stages 2, 3, 4 and 5 were non-acidfast. These conclusions were believed to represent an integration of several aspects of dissociation in the tubercle bacillus, portions of which had been reported by Mellon and by other investigators over a course of many years. One important consequent of this study is to show the futility of attempting to demonstrate either the presence or absence of a bacterial life cycle through the limited range of observations made possible through a brief series of stained microscopic, or even microcultural, preparations.

Marchal[100] in a voluminous monograph published in 1932 and entitled "Variation et Mutation en Bacteriologie," undertook to present a zoologist's viewpoint on bacterial variation, as examined with special reference to B. coli mutabile and several chromogenic species. The cultural attributes serving this study were chiefly chromogenesis and fermentation reactions, while morphological characters of cell and colony received scant attention. The author obseved that variation might be either temporary or hereditary. Among the hereditary variants he noted that some might be due to the gain of a new character or to the revival of one that had become latent. He proposed the use of the term, mutation, for any hereditary modifications appearing suddenly, "since de Vries did not limit the use of the term, mutation, to organisms reproducing by a sexual process." He found it difficult to speak of microbic dissociation because it was based on characters so contingent as smooth or rough, flat or raised colonies, features which he could produce easily by modifying his mediums. He was struck by the circumstance that the bacterial cell, which appeared rather simple when viewed for the first time, was in reality a fairly complex thing; and admitted the difficulty of comparing variation features in different strains. In conclusion, it seemed easier to agree with van Loghem,—that it will be only when we recognize variability as being a function of the individual cell, and when we have separated the phenotypes from the genotypes, that it will be possible to contribute effectively to genetics through the observations in microbiology. One may agree with Marchal that, for the present at least, geneticists will receive little help from knowledge of bacterial variation; and it is not apparent that bacteriologists will get much help from the geneticists. After all, these issues will probably have to be solved by the bacteriologists themselves.

Lewis,[88] after studying the phenomenon of secondary colony formation and its probable cause and significance in B. mycoides, concluded that the explanation having the most in its favor was Neisser's theory of variation and adaptation to nutrient materials; and that there was no evidence in mycoides that secondary colonies "represent a special phase in a complex, pleomorphic life history through which the organisms must pass in a cyclogenetic manner of development." The clear recognition of the distinct phasic nature of secondary colony formation is often a difficult matter, and it seems probable that the technical approach used by Lewis was not adequate to render decipherable certain important aspects of the phenomenon that have clearly been read by others who have in the past studied corresponding colony features in the same and other species.

Wycoff[209] from his motion pictures of several bacterial species concluded that the most important factor contributing to pleomorphism "may be defined by saying: (1) that bacteria can divide in more than one way, and (2) that the rate of division can vary independently of the rate of growth," both being influenced by the kind and richness of the medium. This is probably the simplest theory of pleomorphism that has been propounded. If the observed extent of cellular pleomorphism were limited to cocci, short rods, long rods and filaments, and if there were no tendency for these elements to become stabilized in certain culture phases possessing widely different characteristics, the suggestion might possess merit. While branching rod forms offered some difficulty in interpretation on the basis of the variation principles advanced, most of the "swollen, distorted and micrococcoid cells . . . have the appearance of being injured organisms—"and thus were conveniently removed from elements requiring interpretation. While Wycoff has contributed significantly to existing knowledge of cell division and changing cell morphology through his moving picture records, it is questionable whether (considering the present limitations of the method for long continued observations under favorable cultural conditions) such methods offer at present sufficient grounds for far-reaching conclusions on either the presence or absence of life cycles in bacteria, or on the significance of pleomorphism.

Colien,[18] working with a pigment producing coccus, recently reported the elements of a pleomorphic cycle involving chiefly the coccus and a filamentous form. The former transformed into rods and filaments, the latter subsequently reverting to cocci. There was also established a certain directional trend, apparently possessing the significance of a simple cycle, although the time element was rather short. The author also reported that a filtrable form played a part in this development.

From this brief survey of views currently held regarding the biological significance of dissociative variation it appears that there exists little unanimity of opinion. Although the facts of dissociative behavior have now been accepted by the majority of bacteriologists, their biological significance is commonly regarded as still unrecognized; and a few bacteriologists imply that dissociative phenomena possess no significance other than the results of environmental influences. The latter views ordinarily resort to the "injury hypothesis," and probably unconsciously summon Lamarkian principles to assist in bridging the gap of the necessarily involved hereditary transmission of the acquired modifications. The facts of genetics, however, commonly present no obstacles to bacteriologists.

So far as life cycles are concerned, necessarily embracing a conception of ontogenesis in bacterial development—when such conceptions do not awaken fears of grave injury to the science of bacteriology as a result of tampering with such notions, few except those who have devoted much time and thought to this possibility are inclined to admit the existence of cycles. Opponents of this view seem still to believe that life cycles among the bacteria are necessarily "complex" or "complicated," whereas in reality some are surprisingly simple. Haag's[63] cycle of anthrax species and Mellon's[113] cycle of the tubercle bacillus can scarcely be termed complicated. And the same is true in other instances.

The situation regarding these matters may be summed up by saying that, although the monomorphic conception has largely been abandoned, no completely satisfactory explanation of the facts of pleomorphism in general, or of the dissociative variants in particular has been substituted. For this reason it appears to many bacteriologists that the phenomena of variation have suddenly become more numerous and complicated without becoming any more clear. But it is probably more equitable to the science of bacteriology to accept well-supported facts, though not understood, than to deny them for the same reason. I believe, moreover that, there exists a satisfactory explanation for the existence of that category of variants observed in dissociative variation; and this view I wish to outline in the following section.

A viewpoint on the variation problem.—While a satisfactory explanation of the intricate manifestations of variability among the bacteria will probably not be attainable for a long time, my own observations and experiments, covering a period of many years, together with such an appraisal of the works of others as has been attempted in this and my earlier review, has led me to a belief that at least a partial understanding of the problem may be attained more quickly if some attempt is made to disengage from each other the several categories of variability that appear to be operative in the variation phenomena of many bacterial species. It seems to me that the greatest difficulty obstructing clearer insight is due to the circumstance that variation phenomena, broadly considered, are due to the operation of at least three different categories which can act concurrently or separately. These may be designated as follows: (1) ontogenetic variations; (2) impressed variations; and (3) fluctuating variations in the Darwinian sense. Of these the first would embrace the chief culture phases, together with such phases as have not yet been recognized. The second would include all of those modifications determined by the pressure of environment; and these environmental influences may be regarded as operating on any one or on all of the ontogenetic variants of the first category.* The third category would comprise all those random fluctua-

---

* And, since the first category of variation is conceived as embracing the chief culture phases (M, S, R, and possibly G), it might be important eventually to ascertain on which culture phase

tions so characteristic of all living things, and for the origin of which, even in the higher forms, no satisfactory explanation has been offered either by Darwin or his followers. To what extent, if at all, nuclear reorganizations may enter into the production of this aspect of the variation picture in bacteria is still unknown, although Lindgren[89a] has attempted to supply evidences for the existence of a cytological and nuclear basis for such reorganizations. Unquestionably such studies are highly desirable, but one cannot fail to be impressed with the incongruity of the circumstance that a geneticist has advanced to the point of depicting gene strings in bacteria before the majority of bacteriologists themselves have even come to accept the existence of a bacterial nucleus.

In the following section, however, no attempt will be made to analyze the variation problem further with reference to the categories 2 and 3. What follows will be related chiefly to the first category, the phase variants; and an attempt will be made to demonstrate that, as stages in the gradual development of the species, they should be removed from the general field of variability as this term is commonly understood.

The first category of variability would be represented by those variants that have received special attention in studies on microbic dissociation; in other words, they would include the dissociative variants or culture phases, such as the mucoid, the smooth and the rough, which are already fairly well known. These forms of culture give considerable evidence of being concerned with the ontogenetic development of a bacterial species.[*]

Before attempting to segregate those instances of so-called variability that present this evidence of being concerned with the ontogeny of the species, it is important to have as clearly as possible in mind what in reality constitutes a bacterial species as this may be interpreted in the light of recent studies on dissociative variation. Since the species-concept has never been clearly or satisfactorily defined in the entire history of bacteriology, I am quite aware that an attempt to do so at this late date borders on rashness; and especially since the suggestion I have to make is foreign to all current lines of bacteriological thought—although not so foreign to certain conceptions regarding the nature of species in some of the higher bacteria, entertained by a small group of European bacteriologists and botanists during the latter years of the past century.[†]

---

environmental influences are able to exert the greatest pressure in order to effect modifications. At present it might appear that the smooth phase culture is much more sensitive to environment than the rough; and that is probably the basis of the common view that the smooth is "less stable."

[*] While it must be conceded that the evidence is not yet sufficient definitely to place the phase variants as ontogenetic stages in the development of the species-microphyte, the accumulated data bear strongly on this side of the question. I am confident that, in the course of time, the demonstration will be complete; and in the meantime, for purposes of argument, this assumption will be made in the accompanying presentation.

[†] I refer here to the older group of morphologists including Lancaster, Cienowski, Zopf, Vicentini, Billet and others. It was in connection with the purely morphological studies of this early group that the terms, "dissociation" and "culture phase" arose and were first commonly

Let us start by inquiring: What, for example, is the typhoid species? And by what is it typified? Current bacteriology is inclined to answer the second question only: This species is typified by the bacillus of Eberth. But what then is the "bacillus of Eberth"? The reply, presumably taken from Bergey's "Manual," is sufficiently prompt and very concise: The bacillus of Eberth, correctly known as Eberthella typhi, is a rod-shaped organism belonging to the Bacteriaceae, and measuring about so-and-so and fermenting this-and-that . . . and producing typhoid fever in man. In other words, the typhoid species is typified by the typhoid bacillus; and the characterization of this "species," in order that there may be no mistake about recognizing it, is set down in a page of type.

But, if this be true and the typhoid species is conceived of as so narrowly-limited an entity, what shall be done with the remaining culture forms which are every bit as "normal," but which do not conform to the classical specific requirements either in cell form, colony structure, biochemical reactions, serological reactions or virulence? There was a time when this difficulty could be solved by the simple affirmation that these other forms "did not belong." In more recent years, however, it has been learned that they do belong, but that the systematists do not know what to do with them. If they could be disposed of as abnormal or illegitimate variants left on the front doorstep of the bacteriologist by an unkind fate, the problem could be settled happily, for the bacteriologist at least, by their relegation to that asylum for unwanted variants, the "involution" category. But there, too, the signs are now up discouraging further admissions. Moreover, the illegitimates seem about as physically fit for bacterial existence as those other forms happily born to the species. They are, moreover, beginning to put up a considerable clamor to be given a decent treatment by those unwilling earlier to adopt them. In other words, in no manual, textbook, or journal can one discover a definition or adequate characterization of the typhoid species. Or, for that matter, of any other.

Many of these difficulties could be settled very equitably for both bacteria and bacteriologist if it could be proved that certain things relating to these variants were true; and there is increasing reason to believe that they are. The necessary adjustment, which would presumably come as a shock to the majority of conservative bacteriologists, involves a recognition of the fact that what is commonly known as the typhoid bacillus (to continue the example) does not in any sense, in and by itself, typify the species, Eberthella typhi; nor does the characterization given to Eberthella typhi describe the typhoid species. On the contrary, it is becoming increasingly clear that this species, regarded in its entirety, is typified by a series of morphologic entities which, taken collectively and perhaps in a definite sequence, combine to present the picture of an organized living thing—the typhoid species-microphyte, an organized botanical entity. Such a microphyte may therefore be conceived of as comprising, within the full range of its development (ontogeny), each and every culture phase, and every cell form characterizing those phases. Under these conditions the culture phases (mucoid, smooth, rough, etc.) assume the new significance of stages in the course of the development of the species-microphyte (ontogeny), and not that of variants within the common meaning of that term.

To this it may be added it is reasonable to expect that each ontogenetic stage (culture phase) is susceptible to the influence of both contingent (environmental) and inherent (nuclear) variation factors—just as if each phase were, in itself, a complete biologic entity. What I mean by this is that the individual cells of every culture phase (and of intermediate phases as well) are susceptible to the influence of environment; and perhaps, at certain points in the ontogenetic process (as in the rough phase cultures), to variation factors involving some degree of nuclear reorganization.

But the question may pertinently be asked—How can one conceive of such a species-microphyte when one is dealing with bacterial cells manifesting so marked a tendency for individual and detached existence as is commonly observed in all the culture phases except the rough, in which the organisms enter into chain formation or filamentous structures, often represented by a tangled mycelium as in the bacillus of Friedländer, the influenza bacillus and other species? To accomplish such a visualization of a bacterial species among the lower bacteria is presumably difficult for those not accustomed to thinking along these lines. But it is not impossible for those who are sufficiently open minded and who can develop a botanical viewpoint on problems of bacterial morphology. It was possible for the older group of morphologists, as previously pointed out; and it may be noted that it was these early workers who first made use of the term, "dissociation," to indicate the detachment of individual cell entities from the species-microphyte which, in the material then studied, was most commonly one of the higher bacteria; also the term, "phase" which was applied to any one of the detached organisms, multiplying for a time independently, as in the case of the coccus and bacillary forms of Leptothrix racemosa and buccalis. Indeed four "phases" in the life history of Leptothrix were pointed out by Vicentini[197] many years ago.

While the early application of the dissociation concept could not be carried out to the degree implied, for example, by those still earlier morphologists who attempted to recognize in Leptothrix the common ancestor of a highly diverse bacterial progeny, the fundamental viewpoint in these early surmises was not so seriously in error as many modern bacteriologists believe. Indeed, it is probable that this earlier conception of species-dissociation or fragmentation, when brought within limits, and applied to a specific bacterial population, observed over a considerable time, is the explanation of much that to many is still obscure in the field of bacterial variation; and particularly in dissociative variation as this province has been developed within recent times.

But, returning to the question of visualizing the species-microphyte existing under conditions in which the individual cells are detached and scattered, as in the smooth phase cultures. In species of the higher bacteria and the fungi it is easier to picture the microphyte as an organized plant unit, although here also (particularly in the higher bacteria) certain cell components may carry on a detached existence. The reason for this is that, in the "higher" forms, the cell organization is more coherent in the progress of the specific cyclogeny; and, aside from the formation and liberation of conidia, the detachment of individual cells from the species-microphyte does not occur so commonly as in bacterial species. But, in the fungi, we do not term a single hypha the fungus; nor do we term a conidium, nor a conidiophore, nor a sterigma, the fungus. We recognize them as differentiated cellular elements, each possessing a characteristic form and probably a separate function. But they must be in combination in order to present an adequate picture of the entire fungus "plant," the microphyte. If, indeed, all of the cellular structures that characterize the fungus organization should become separated and develop individually for a time, but still carrying on the ontogenetic scheme, it is possible that bacteriologists, or even mycologists, might have considerable difficulty in putting them together again in their proper developmental sequence. And perhaps that indicates the reason why it seems so difficult to reassemble the culture phases (mucoid, smooth, rough, etc.,) of bacterial species in order to produce out of them the bacterial species-microphyte—which must be pictured by imagination rather than actually observed as a unit thing.

To me it is not difficult to picture the bacterial species in this manner. The isolated and scattered cells (aside from those of the rough phase) may be regarded merely as single elements that are detached from the microphyte organization, to which they however still belong, and in whose complete structure they still have a place. While they appear to be developing independently, free from all relations and control, each may be considered as helping to maintain the developmental trend characteristic of the species-microphyte. In other words, the specific bacterial entity is not the cell, nor any colony of cells, but the microphyte, which is a cell-composite embracing them all in a more or less continuous development. In this way the species-microphyte comprises all of the stages present in its own ontogeny; and these stages are the culture phases that are now so commonly recognized. It may appear, moreover, that the developmental trend manifested in some instances, as for example the $M \rightarrow S \rightarrow R$, is in some way related to the cycle, in which the rough phase culture appears to represent the culmination of progressive, cytological development; and from which (probably through the liberation of gonidia) the new cycle starts.

In concluding this section I may add that I regard the foregoing conception of the biological significance of dissociative variation, formulated on the basis of evidence assembled largely during the past fifteen years, as representing an inviting, but still unproved, hypothesis. There exist many facts in its favor; and some not in its favor. What will become of it will depend upon the extent to which bacteriologists who are still interested in problems of the pure science are able to focus their investigations upon the links that are still missing from the chain of evidence. Of these I regard as most important the following: (1) Establishing for as many species as possible what may be termed the chief culture phases; (2) establishing the natural trend in the inter-phase transformations; (3) ascertaining the nature of the cytological mechanism involved in these transformations; (4) demonstrating the part played by the mature rough phase culture in the formation and liberation of gonidia; (5) demonstrating the development of the gonidia into the succeeding culture phase.

If the hypothesis presented in the foregoing pages possesses any elements of truth their bearing on a point mentioned on a earlier page now perhaps becomes clear: Namely, that the dissociative variants, regarded as stages in an ontogenetic process, can not justifiably be allied to those variants which depend on other variability factors such as influence of environment and a possible process of nuclear reorganization. My meaning regarding the logical segregation of the culture phases from the general field of variability may perhaps be made more clear by the following example: Whether a bacterial culture appears as mucoid, smooth or rough, depends upon the factors that control ontogenetic development, as in all plant and animal species, and not on variability factors in the ordinary sense. But, whether any one of these culture phases (in a species ordinarily non-chromogenic) becomes red or yellow or pink, depends upon other influences—perhaps those of environment, perhaps of nuclear reorganization. In future studies on bacterial variation (in the wider sense) it will be of importance to ascertain what sort of attributes are most clearly correlated with culture phase, either positively or negatively; and which appear susceptible of independent variability. While some characteristics, such as chain formation and filamentous structures, are correlated with rough phase in the majority of species already studied, and while there appears to exist a negative correlation between virulence and rough phase in many species, there are likely to be instances in which the order of correlation observed in some species will not be exactly duplicated in others.

BEARING OF DISSOCIATIVE VARIATION ON THE CHARACTERIZATION AND CLASSIFICATION OF BACTERIAL SPECIES

One important issue arising from a survey of the facts presented in this review, as well as of those recorded in my earlier monograph on microbic dissociation,[66] concerns intimately methods and means for accomplishing what is referred to rather inaccurately at present as "the description of bacterial species"; and for classifying them into genera and larger groups, as attempted by Bergey[8] in his manual. The scheme that has been employed for this purpose since the beginnings of bacteriology, and which has been continued by Bergey to produce his considerable list of species characterizations, is firmly based on the old monomorphic conception of the nature and organization of bacteria. For each bacterial species there is described a single form (phase) as if this typified the entire species. The culture phase described may be any one of three that are now clearly recognized; but it is likely to be the one most common to bacteriologists, or perhaps one that is regarded as typical because it seems to agree with a culture described many years ago at the time when the "species" was first named. Indeed, it is apparent that modern recognition of distinct culture phases has not yet been permitted by the systematists to interfere in any appreciable way with the problems and methods of taxonomy.

Of course the chief obstacle in the path of developing a logical, adequate and scientifically exact classification of bacteria is that systematic bacteriologists, possessing no tangible conception of what a bacterial species actually is, experience some difficulty in ascertaining just what it is that should be classified. Indeed, the present situation in systematic bacteriology is such as to well merit the question raised by Zinsser and Bayne-Jones[215] when they inquire—"Are the bacteria classifiable?". And I am strongly convinced the correct answer is that they are not—at least by any such methods, or on such a basis, as those employed by Bergey and his collaborators. I believe, moreover, that the bacteria will not be classifiable in any real sense until bacteriologists have succeeded in learning what it is they have to classify. In other words, until they know what comprises a bacterial species. The entire so-called classification of bacteria as it exists at the present moment is nothing more than an attempted cataloging of species-fragments. It is helpful if one wishes to compare fragments but this should not be mistaken for a comparison of species. Indeed, what little order has already been brought into species-differentiation and classification has not been due to any logical approach to the taxonomic problems involved.

In order to be sufficiently specific in these criticisms, it may further be pointed out that, even if the taxonomists were warranted in limiting their species characterizations to a single culture phase, there is no rule applicable to different species, indicating what phase, among the several that are now recognized, should be regarded as most "typical." The phase chosen might be the smooth, as in the typhoid species, the rough as in the anthrax species, or the mucoid as in the pneumococcus and Friedländer's bacillus. How can one present an adequate characterization of the entire species of these or of any other micro-organism while excluding from his descriptions two-thirds to three-quarters of the observed facts appertaining to their essential and normal attributes? Again it may be emphasized that a species is not typified by one phase alone, but by the summation of all its phases together with their very individual attributes.

In concluding this section on the significance of dissociative variation, which is presented with no other purpose than to awaken some recognition on the part of bacteriologists regarding the inadequacy of the present species-concept and, pari passu, of currently employed principles of bacterial classification, I wish to add the following: If the facts of microbic dissociation, as indicated in this and my earlier review, have accomplished any one thing of importance for the science of bacteriology, it is the demonstration that a bacterial species, regarded in its entirety, is a very complex thing; that it is made up of culture forms which, from one moment to another, can assume the most diverse characters—but without, so far as now known, transcending the actual species limits. Furthermore, that these diverse forms (aside from variability factors of which we still know little) represent the so-called culture phases. It has, moreover, been amply demonstrated that these phases run a more or less parallel course with respect to their appearance, trends and probable functions in a considerable number of bacterial species. And, finally, that they are all strictly normal manifestations of bacterial development. That these phases, which provide the basis of dissociative variation, pursue a definite trend in the course of their intertransformations appears probable. But, whether they do or do not, has no bearing on the legitimacy of their incorporation into any scheme of species characterization and classification. The circumstance that they exist as normal species components is sufficient to justify their inclusion. And it should be added that no characterization of the species can be complete without all of them—for they are the species.

GENERAL CONCLUSIONS

The foregoing account of the many investigations presented in this review relating to dissociative variation has a particulate significance for many special problems. But it offers, at the same time, the vantage point for a broader outlook upon the nature and scope of what is commonly termed "the science of bacteriology."

This science, compared with some others such as chemistry and physics for example, manifests several peculiarities, clearly recognized by those who have felt that, despite its important contributions to human welfare, there has "been something wrong about bacteriology" as a science. It appears possible to define some of these peculiar characteristics. First, there is probably no other science in which the "pure" scientific aspects manifest so small and undeveloped a structure in proportion to its valuable applications. Second, it is probable that in no science so much as in this, have there been performed so many experiments and investigations the net results of which have been of such slight importance to the mother science. Third, there is probably no science in which so large a proportion of the workers have so small in interest in the fundamental problems of the pure science. This strange circumstance, apparent to all who give serious consideration to the broader problems of bacteriology, must have an explanation; and I believe it may be found in the following facts.

In the first place, pure bacteriology must be regarded as a still relatively undeveloped subject. This is due in part to restrictions placed upon its progress by the demands of the monomorphic conception many years ago. It is also due to a considerable extent to the circumstance that bacteriology has, since the early days of the morphologists, been submerged in a maze of attempted applications. This has resulted in side-tracking investigators from the less profitable, and considerably more dull, problems of consequence to the pure science. In this connection, one may agree with Topley and Wilson[188] when they write: "It is almost true to say that the present position of bacteriology is due to the fact that there have been no bacteriologists. From Pasteur onward, bacteriologists have been more interested in what bacteria do than in what they are; and much more interested in the ways in which they interfere with man's health and pursuits than in the ways in which they function as living beings."

In the second place, there has existed, in the rather small parcel of knowledge represented by the subject-content of bacteriology, no integrating principle. Bacteriology has had no "laws" resembling in any way those established long ago for other sciences. When one speaks of the "laws" or "principles" of bacteriology, close examination shows that these laws are the laws of physics and of chemistry, or of physiology. It is probably for this reason that so large a proportion of contributions to bacteriology have nothing to tie to; no background of enduring fabric in which the observations can be integrated, either with the effect of supporting a principle already established, or of creating a new one. It is also because of this situation that some have even been inclined to question whether bacteriology is entitled to the rank of science.

Some years ago, James Truslow Adams in his entertaining biography of the Adams family, remarked: "I see no use and much disadvantage in calling anything a science which has not so succeeded in arranging its facts and observations as to have disclosed 'laws,'—that is, uniform modes of behavior which show regular recurrence and thus possess predictability." While many may not accept this criterion of what is entitled to rank as a science, the quotation given above reflects with considerable faithfulness a situation that has existed in bacteriology since its beginnings. Indeed, bacteriology has "no laws," and almost "no bacteriologists." These lacks, which so frequently seem to be accepted as a matter of course, measure the handicaps under which this science has labored and developed.

But what is the bearing of all this on the newer studies on dissociative variation? In the first place it is apparent that, not since earlier days when morphological studies were paramount, have so many bacteriologists applied themselves to problems of pure bacteriology as during the past decade; for these workers have in considerable number attempted to expand the knowledge of microbic dissociation; and the most fundamental problems in this field are problems in pure bacteriology. Second, it is unquestionably true that a greater body of facts relating to the nature, organization and behavior of bacteria has been acquired during the past fifteen years than had been gained during the previous half century of endeavor under the older viewpoint. Finally, the most important, it appears that the results obtained in studying the biology of bacteria from the new viewpoint, freed from all the inhibiting influences of the monomorphic conception, have made possible, for the first time in the history of bacteriology, the development of certain integrating principles that soon may be introduced into the science. In other words, it is apparent that there are beginning to evolve some of those important facts upon which the first definite "laws of bacteriology" may be based; and from the firm establishment of which bacteriology may sometime emerge to take its place as an actual, rather than as an implied science.

LITERATURE CITED

1. Alloway: Cited by Gay.[54]
2. Almeden: Thesis, University of Pittsburgh, 1936.
3. Almon and Baldwin: J. Bact. **26**: 229, 1933.
4. Andrewes: J. Path. & Bact. **31**: 132, 1928.
5. Andrewes and Christie: The Haemolytic Streptococci, Medical Research Council, Special Report series No. 169, London, His Majesty's Stationery Office, 1932.
6. Arkwright: Extract from First Internat. Cong. of Microbiol., Paris, 1930.
7. Arkwright: Bacterial Variability, System of Bacteriology, **1**: 1930.
8. Bergey: Manual of Determinative Bacteriology, Baltimore, Williams and Wilkins, 1931.
9. Birkhaug: Ann. Inst. Pasteur **54**: 1935.

10. Bitter: Deutsche med. Wchnschr. 1298, 1922.

11. Blake and Trask: J. Bact. **25:** 289, 1933.

12. Brueckner and Sherman: J. Infect. Dis. **51:** 1, 1932.

13. Buchaley: Centralbl. f. Bakt. (Abt. I) O., **119:** 444, 1931.

14. Burnet: J. Path. & Bact. **32:** 15, 1929.

15. Cantacuzene: Compt. rend. Soc. de biol. **99:** 1003, 1932.

16. Carpenter and Long: J. Bact. **25:** 241, 1933.

17. Christison: J. Path. and Bact. **36:** 285, 1933.

18. Colien: J. Bact. **30:** 301, 1935.

19. Cooley: J. Infect. Dis. **50:** 358, 1932.

20. Cowan: Brit. J. Exper. Path. **8:** 6, 1927.

21. Craig and Johns: Proc. Exper. Biol. & Med. **29:** 5, 1932.

22. Cunningham: Centralbl. f. Bakt. **82:** 25, 1930; **82:** 481, 1931; **83:** 22, 1931; **83:** 1, 1931; **83:** 220, 1931.

23. Dahr and Kolb: Deutsche med. Wchnschr. **61:** 1871, 1935.

24. Davis: J. A. M. A. **58:** 1852, 1912.

25. Dawson: J. Exper. Med. **51:** 123, 1930.

26. Dawson: J. Path. & Bact. **36:** 263, 1933.

27. Dawson: Proc. Soc. Exper. Biol. & Med. **31:** 590, 1934.

28. Dawson and Olmstead: Science **80:** 296, 1934.

29. Delves: J. Infect. Dis. **50:** 350, 1932.

30. Dienes: J. Infect. Dis. **57:** 12, and 22, 1935.

31. Dienst: J. Bact. **26:** 489, 1933.

32. Dochez, Mills and Kneeland: Proc. Soc. Exper. Biol. & Med. **52:** 314, 1932.

33. Dudgeon and Durru: J. Hyg. **32:** 275, 1932.

34. Dulaney: J. Infect. Dis. **42:** 575, 1928.

35. Duval and Luzenberg: Proc. Soc. Exper. Biol. & Med. **30:** 272, 1932.

36. Eagles: Brit. J. Exper. Path. **9:** 330, 1928.

37. Eaton: J. Bact. **24:** 271, 1934.

38. Edwards: J. Bact. **24:** 283, 1932.

39. Emmons: Arch. Dermat. & Syph. **25:** 987, 1932.

40. Enders: J. Bact. **23:** 83, 1932.

41. Ettinger-Tulcyznska: Ztschr. f. Hyg. u. Infektionskr. **114:** 769, 1933.

42. Evans: J. Infect Dis. **52:** 39, 1933.

43. Evans: Pub. Health Rpt. **47:** 1723, 1932.

44. Fabian and McCullough: J. Bact. **27:** 583, 1934.

45. Falk, Harrison, McKinney and Stuppy: Am. J. Hyg. Monographic Series, No. 11, 1931.

46. Fejgin: Compt. rend. Soc. de biol. **106:** 163, 1931.

47. Fei-fang Tang and Castaneda: J. Bact. **16:** 431, 1928.

48. Felix and Pitt: Lancet. **2:** 186, 1934.

49. Felix and Pitt: J. Path. & Bact. **38:** 409, 1934.

50. Felix and Pitt: J. Hyg. **35:** 428, 1935.

51. Felix, Bhatnagar and Pitt: Brit. J. Exper. Path. **15:** 346, 1934.

52. Fleck and Elser: Centralbl. f. Bakt., O. **125:** 180, 1932.

53. Flynn and Rettger: J. Bact. **28:** 1, 1934.

54. Gay and Associates: Agents of Disease and Host Resistance, Springfield, Ill., C. C Thomas, 1935.

55. Gay and Claypole: Arch. Int. Med. **12:** 613 and 621, 1913.

56. Griffith: J. Hyg., 1928, **27:** 113, 1928.

57. Griffith: J. Hyg. **34:** 542, 1935.

58. Grinnell: J. Bact. **19:** 457, 1930.

59. Grinnell: J. Exper. Med. **56:** 907, 1932.

60. Grubb: J. Infect. Dis. **56:** 64, 1935.

61. Grumbach: Centralbl. f. Bakt. (Abt. 1) O. **118:** 206, 1930.

62. Gwatkin: Canad. Pub. Health J. **23:** 485, 1932.

63. Haag: Arch. f. Hyg. **98:** 271, 1927.

64. Hadley, Faith: J. Am. Dent. A. **17**: 1730, 1930.
64a. Hadley, Faith: Thesis, University of Michigan, 1935.
65. Hadley, Bunting and Delves: J. Am. Dent. A. **17**: 2041, 1930.
→ Hadley, Philip: J. Infect. Dis. **40**: 1, 1927.
→ Hadley, Philip; Delves and Klimek: J. Infect. Dis. **48**: 1, 1931.
68. Hadley, Philip: Proc. Inst. Med. Chicago **9**: 1, 1932.
69. Hadley, Philip: J. Bact. **25**: 572, 1933.
70. Hare: Brit. J. Exper. Path. **12**: 261, 1931.
71. Hoffstadt and Youmans: J. Infect. Dis. **51**: 216, 1932.
72. Hoffstadt and Youmans: J. Infect. Dis. **54**: 250, 1934.
73. Holster: Yale J. Biol. & Med. **3**: 321, 1931.
74. Howell and Burton: J. Infect Dis. **55**: 79, 1934.
75. Jennings: Bull. Johns Hopkins Hosp. **49**: 95, 1931.
76. Jordan: Epidemic Influenza, Chicago, American Medical Association Press, 1927.
77. Kahn and Schwartzkopf: J. Bact. **25**: 157, 1933.
78. Kendall: Northwestern Univ. Bull., Nos. 5 and 8, 1931; Science **74**: 129, 1931.
79. Klumpen: Centralbl. f. Bakt. (Abt. I) O., **124**: 241, 1932.
80. Knaysi: J. Bact. **26**: 623, 1935.
81a. Koch and Mellon: Jour. Bact., **19**: 25, 1930.
81. Kopeloff: J. Infect. Dis. **55**: 368, 1934.
82. Koser and Dienst: J. Infect. Dis. **54**: 131, 1934.
83. Koser and Styron: J. Infect. Dis. **47**: 443 and 453, 1930.
84. Kun and Fenyvessy: Centralbl. f. Bakt., **124**: 485, 1932.
85. Lancefield: J. Exper. Med. **59**: 441; **57**: 571, 1934.
86. Leslie and Gardiner: J. Hyg. **31**: 423, 1931.
87. Lewis: J. Bact. **24**: 381, 1932.
88. Lewis: J. Bact. **25**: 359, 1933.
89. Li, C. P.: J. Exper. Med. **30**: 40, 1930.
89a. Lindgren: Proc. Exper. Biol. & Med. **30**: 83, 1932.
89b. Locke: J. Bact. **31**: 82, 1936.
90. Loewenthal: Ztschr. f. Hyg. u. Infektionskr. **113**: 445, 1932.
91. Loewenthal: Ztschr. f. Hyg. u. Infektionskr. **114**: 379, 1932.
92. Loewenthal: Brit. J. Exper. Path. **64**: 403, 1934.
93. Long and Bliss: J. Exper. Med. **60**: 619, 1934.
94. Lyons: J. A. M. A. **105**: 1972, 1935.
95. Lyons and Ward: J. Exper. Med. **61**: 531, 1935.
96. Maegraith: Brit. J. Exper. Path. **14**: 227, 1933.
97. Mallman and Gallo: J. Agric. Research **46**: 267, 1933.
98. Maltaner: Jour. Immunol. **26**: 161, 1934.
99. Marassini: Centralbl. f. Bakt., O., **71**: 113, 1913.
100. Marchal: Variation et Mutation en Bacteriologie, Libraire français, Paris, 1932.
101. Marshall and Jared: J. Infect. Dis. **49**: 318, 1931.
102. McDaniels and Neal: Proc. Soc. Exper. Biol. & Med. **30**: 115, 1932.
103. McGaughey: J. Path. & Bact. **36**: 263, 1933.
104. McKenzie, Fitzgerald and Irons: J. Bact. **29**: 583, 1935.
105. McKinney: J. Bact., 1933, **27**: 373, 1933.
106. Mellon: Proc. Soc. Exper. Biol. & Med., in press.
107. Mellon: Proc. Soc. Exper. Biol. & Med., 1931, **29**: 206, 1931.
108. Mellon: In Goldberg: Clinical Tuberculosis, Philadelphia, F. A. Davis Co., **1**: Chapter 1935.
109. Mellon, Almeden and Richardson: J. Bact. **31**: 75, 1936.
110. Mellon and Fisher: J. Infect. Dis. **51**: 117, 1932.
111. Mellon and Fisher: J. Bact. **23**: 18, 1932.
112. Mellon and Jost: Am. Rev. Tuberc. **19**: 483, 1929.
113. Mellon, Richardson and Fisher: Proc. Soc. Exper. Biol. & Med. **30**: 90. 1932.
114. Meyn: Thesis, Univ. of Leipzig, 1930.
115. Morton: Thesis, Univ. of Mich., 1936.

116. Nelson: J. Exper. Med. **45**: 389, 1927.
117. Neisser: Centralbl. f. Bakt. (Abt. I), O., **124**: 503, 1932.
118. Neufeld and Levinthal: Ztchsr. f. Hyg, u. Infektionskr., **55**: 324, 1928.
119. Nichols: J. Bact. **30**: 445, 1935.
120. Ninni: Compt. rend. Soc. de biol. **107**: 615, 1931.
1: ➞ Nungester: J. Infect. Dis. **44**: 73, 1929.
122. Nungester: J. Infect. Dis. **45**: 214, 1929.
123. Nungester and Junge: Proc. Soc. Exper. Biol. & Med., **28**: 681, 1932.
124. Oesterle and Stahl: Centralbl. f. Bakt. (Abt. 2), **79**: 1, 1929.
125. O'Neil: J. Bact. **26**: 521, 1933.
126. Oppenheim: Centralbl. f. Bakt. (Abt. 1), O., **111**: 83, 1929.
126a. Orr, Josephson, Baker and Redd: Canad. J. Research **9**: 350, 1933.
127. Panek and Zakharoff: Compt. rend. Soc. de biol. **104**: 607, 1930.
128. Paul: J. Bact. **28**: 45, 1934.
129. Paul: J. Bact. **28**: 69, 1934.
130. Petroff: J. Exper. Med. **51**: 831, 1930.
131. Petroff, Branch and Steenkin: J. Bact. **17**: 58, 1929.
132. Petroff and Steenkin: J. Exper. Med. **51**: 831, 1930.
133. Petroff and Steenkin: J. Infect. Dis. **56**: 277, 1935.
134. Pilot: J. Infect. Dis. **55**: 228, 1934.
135. Pilot and Davis: J. A. M. A. **97**: 1691, 1931.
136. Pinner and Voldrich: Am. Rev. Tuberc. **24**: 73, 1931.
137. Pirie: South African M. J. **4**: 191, 1929.
138. Pittman: J. Exper. Med. **53**: 471, 1931.
139. Pittman: J. Exper. Med. **58**: 683, 1933.
140. Plastridge and McAlpine: J. Infect. Dis. **46**: 315, 1930.
141. Plastridge: J. Infect. Dis. **50**: 146, 1932.
142. Pokrowskaja: Centralbl. f. Bakt. (Abt. 1), O., **116**: 304, 1930.
143. Prissick: Canad. J. Research 1933, **8**: 217, 1933.
144. Rake: J. Exper. Med. **57**: 549, 1933.
145. Raven: J. Infect. Dis. **55**: 328, 1934.
146. Reed: J. Bact., **31**: 32, 1936.
147. Reed and Rice: Canad. J. Research **5**: 111, 1931.
148. Reed and Orr: Science **76**: 372, 1932.
149. Reimann: J. Exper. Med. **45**: 807, 1927.
150. Reimann: J. Exper. Med. **49**: 237, 1929.
151. Reimann: J. Clin. Investigation **14**: 311, 1935.
152. Reimann: J. Bact., in press.
153. Rettger and Gillespie: J. Bact. **26**: 289, 1933.
154. Rettger and Gillespie: J. Bact. **30**: 213, 1935.
155. Richardson and Mellon: Proc. Soc. Exper. Biol. & Med. **29**: 451, 1932.
156. Roe: J. Bact. **27**: 46, 1934.
157. Rosenow: J. Infec. Dis. **48**: 304, 1931.
158. Rosenow: Proc. of the Staff Meetings of the Mayo Clinic **10**: 410, 1935.
159. Roth: Thesis, University of Michigan, 1933.
160. Sanarelli and Allesandrini: Compt. rend. Soc. de biol. **104**: 1241, 1930.
161. Sanarelli and Allesandrini: Compt. rend. Soc. de biol. **106**: 246, 1931.
162. Sanarelli and Allesandrini: Compt. rend. Soc. de biol. **108**: 405 and 460, 1931.
163. Seastone: J. Bact. **28**: 481, 1934.
164. Seastone and Lawrence: J. Infect. Dis. **52**: 20, 1933.
165. Sergent and Pribriano: Compt. rend. Soc. de biol. **100**: 546, 1929.
166. Sherman and Safford: Science **73**: 448, 1931.
167. Sherman, Safford and Brueckner: Proc. Internat. Dairy Conference, Copenhagen, 1931.
168. Shibley and Haelscher: J. Exper. Med. **64**: 403, 1932.
169. Shimidsu: Centralbl. f. Bakt. (Abt. 1), O., **71**: 338, 1913.
170. Shinn: Thesis, University of Pittsburgh, 1936.
1 ➞ Smith. Muriel: J. Hyg. **31**: 321. 1931.

172. Smith, Sano and Jarema: J. Infect. Dis. **55**: 105, 1934.
173. Smithburn: J. Exper. Med. **61**: 395, 1935.
174. Smithburn: J. Exper. Med. **62**: 645, 1935.
175. Spicer: J. Bact. **26**: 505, 1933.
176. Steenkin: J. Infect. Dis. **56**: 273, 1935.
177. Steenkin, Oatway and Petroff: J. Exper. Med. **60**: 515, 1934.
178. Stevens: J. Infect. Dis. **57**: 275, 1935.
179. Stoughton: Proc. Roy. Soc., London, s. B. **105**: 469, 1929.
180. Stoughton: Proc. Roy. Soc., London, s.B. **111**: 46, 1932.
181. Soule: J. Bact. **23**: 30, 1932.
182. Swingle: J. Bact. **29**: 467, 1935.
183. Swingle: Proc. Soc. Exper. Biol. & Med. **31**: 891, 1934.
184. Swingle and Koser: Proc. Soc. Exper. Biol. & Med. **30**: 1235, 1933.
185. Todd: Brit. J. Exper. Path. **8**: 289, 1927.
186. Todd: Brit. J. Exper. Path. **9**: 1, 1928.
187. Todd and Lancefield: J. Exper. Med. **48**: 751, 1928.
188. Topley and Wilson: Principles of Bacteriology and Immunity, New York, William Wood and Co. 1929.
188a. Tunnicliff: J.A.M.A. **74**: 1386, 1920.
188b. Tunnicliff: J. Infect. Dis. **45**: 235, 1929.
189. Tunnicliff: J. Infect. Dis. **49**: 357, 1931.
190. Tunnicliff: J. Infect. Dis. **48**: 511 and 537, 1931.
191. Tunnicliff: J. Infect. Dis. **52**: 39, 1933.
192. Tunnicliff: J. Infect. Dis. **53**: 280, 1933.
192a. Tunnicliff: J. Bact. **29**: 37, 1935.
19. → Tunnicliff: J. Infect. Dis. **58**: 92, 1936.
193. Tunnicliff and Woolsey: J. Infect. Dis. **56**: 116, 1935.
194. Tzeknovitzer: Ann Inst. Pasteur **45**: 162, 1930.
195. Van Loeghm: Centralbl. f. Bakt. (Abt. 1), O., **83**: 401, 1929; **120**: 318, 1932.
196. Varney and Bronfenbrenner: Proc. Soc. Exper. Biol. & Med., **29**: 7, 1932.
197. Vicentini: Bacteria of the Sputa and Cryptogamin Flora of the Mouth. London, 1897.
197a. Waaler: Studies on the Dissociation of the Dysentery Bacilli, Olso, Jacob Dybwad, 1935.
1' → Walker: J. Infect. Dis. **32**: 287, 1923.
199. Ward and Lyons: J. Exper. Med. **61**: 515, 1935.
200. Wardsworth and Sickles: J. Exper. Med. **45**: 787, 1927.
201. Webster: J. Exper. Med. **45**: 529, 1927.
202. Wheeler: J. Prev. Med. **5**: 181, 1931.
203. White: Further Studies of the Salmonella Group., Medical Research Council, Special Report Series No. 103, London, 1926.
2l → Wilson: J. Hyg. **30**: 40, 1930.
205. Winn and Petroff: J. Exper. Med. **57**: 239, 1933.
206. Winslow: Science **75**: 121, 1932.
207. Winslow: Science 1932, **75**: 121, 1932.
208. Wycoff: J. Exper. Med. **57**: 165, 1933.
209. Wycoff: J. Exper. Med. **59**: 381, 1934.
210. Wycoff: Am. Rev. Tuberc. **29**: 389, 1934.
211. Yasuda: Jap. J. Exper. Med. **11**: 33 and 619, 1933.
211a. Yü: J. Bact. **20**: 107, 1930.
212. Zalatogoroff and Moghelevskia: Compt. rend. Soc. de biol. **99**: 506, 1928.
2 → Zinsser: Science **75**: 256, 1932.
214. Zinsser: Jour. Prev. Med. **6**: 497, 1932.
215. Zinsser and Bayne-Jones: Textbook of Bacteriology, New York Appleton-Century Co., 1934.
216. Personal observation.
217. Personal communication.